CITIES DEMANDING THE EARTH

A New Understanding of the Climate Emergency

Peter J. Taylor, Geoff O'Brien and Phil O'Keefe

First published in Great Britain in 2020 by

Bristol University Press
University of Bristol
1-9 Old Park Hill
Bristol
BS2 8BB
UK
t: +44 (0)117 954 5940
www.bristoluniversitypress.co.uk

North America office:
Bristol University Press
c/o The University of Chicago Press
1427 East 60th Street
Chicago, IL 60637, USA
t: +1 773 702 7700
f: +1 773-702-9756
sales@press.uchicago.edu
www.press.uchicago.edu

© Bristol University Press 2020

British Library Cataloguing in Publication Data
A catalogue record for this book is available from the British Library

Library of Congress Cataloging-in-Publication Data
A catalog record for this book has been requested

ISBN 978-1-5292-1047-7 hardcover
ISBN 978-1-5292-1048-4 paperback
ISBN 978-1-5292-1050-7 ePub
ISBN 978-1-5292-1049-1 ePdf

The rights of Peter J. Taylor, Geoff O'Brien and Phil O'Keefe to be identified as authors of this work has been asserted by them in accordance with the Copyright, Designs and Patents Act 1988.

All rights reserved: no part of this publication may be reproduced, stored in a retrieval system, or transmitted in any form or by any means, electronic, mechanical, photocopying, recording, or otherwise without the prior permission of Bristol University Press.

The statements and opinions contained within this publication are solely those of the authors and not of the University of Bristol or Bristol University Press. The University of Bristol and Bristol University Press disclaim responsibility for any injury to persons or property resulting from any material published in this publication.

Bristol University Press works to counter discrimination on grounds of gender, race, disability, age and sexuality.

Cover design by Blu Inc
Front cover image: Getty
Printed and bound in Great Britain by CPI Group (UK) Ltd, Croydon, CR0 4YY
Bristol University Press uses environmentally responsible print partners

This book is dedicated to grandchildren,
today and in the future, across the world.

Contents

List of Tables and Figures		vi
About the Authors		vii
Preface		ix
1	Declarations: Root and Branch Unthinking	1
2	Alternate: Jane Jacobs' legacy	17
3	Inside Out: Fourteen Antitheses Authenticating Cities	45
4	Reset: Anthropogenic Climate Change Is Urban, not Modern	71
5	Action: Can We Stop Terminal Consumption?	95
References		123
Appendix: Primer on Climate Change Policy		137
Index		145

List of Tables and Figures

Tables

2.1	Jacobs' explosive city growths	25
3.1	Supply and demand: search results from the five IPCC Assessment Reports	47
4.1	Soja's and Ruddiman's time lines	80

Figures

1.1	Anthropogenic climate change without social science	7
2.1	Explosive city growth	24
5.1	The adaptation spectrum	118

About the Authors

Peter Taylor has been a Professor of Geography at Newcastle, Loughborough and Northumbria Universities, and is Emeritus Professor at the latter two. A Fellow of the British Academy and Academy of Social Sciences (UK), with Honorary Doctorates from Oulu and Ghent Universities, he has been honoured by the Association of American Geographers for his research and has been an Advisor to the Chinese Academy of Social Science. He has researched and written widely on political geography and urban studies wherein he is the founder and director of the Globalization and World Cities (GaWC) research network.

Dr Geoff O'Brien, formerly a Senior Lecturer in the Department of Geography at Northumbria University, combines engineering practice with social science work on policy and practice in sustainable development, climate change, disaster management and humanitarian assistance. Prior to joining Northumbria University, he was involved in the geophysical industry with a global remit and a particular focus on environmental responsibility. He served as Lord Mayor of Newcastle upon Tyne in the municipal year of 2012, and was appointed an Alderman of the city in 2019.

Phil O'Keefe, formerly Professor of Environment and Development at Northumbria University, is now an Emeritus Professor. He has worked on large-scale energy and environment programmes, particularly in eastern and southern Africa, that not only accelerated economic growth, but also contributed significantly to the liberation movements, including the anti-apartheid movement. For 30 years he was a lead evaluator of humanitarian assistance throughout the world. He has written extensively on environment and nature with contributions to the understanding of human risk and resilience building.

Preface

There are many books on climate change and cities, so why another one? In this literature cities feature variously as victims (threatened by the rise of sea levels), solutions (compact living) and models (various new 'green' settlements), but always as adjuncts, effectively just add-ons, to the subject of climate change. The basic argument is that given climate change is happening, how are cities affected and what can be done about it?

Acknowledging Hassett's (2017: 14) wide-ranging evidence that cities have made us humans – as she states, 'we haven't just built cities. Cities have built us' – this book offers a completely different take: cities are the key human component in anthropogenic climate change. *Our basic argument is that without cities, there would never have been anthropogenic climate change.* So there – if you want to engage with this unique argument, read on. We think you should because to deal effectively with any emergency – and climate change is now an emergency – it is vital to understand the fundamental mechanism behind the crisis. And we might just be right.

Our book has been a long time in the making. The path to this text begins in 2010 when we three found ourselves together in the same academic home, the Department of Geography at Northumbria University in Newcastle upon Tyne. By coincidence we were in the process of researching and writing two books that were soon to be published: Taylor's *Extraordinary Cities: Millennia of Moral Syndromes, World-Systems and City/State Relations* (Edward Elgar, 2013) and O'Brien and O'Keefe's *Managing Adaptation to Climate Risk: Beyond Fragmented Responses* (Routledge, 2014). This book has its origins in our trying to bring together some of the core ideas of these two texts, the human potentials of urban agglomerations and connectivities, and the building of resilient communities to reduce risks from disruptive events. Although the original texts are very different, they shared common concern for holistic thinking, bottom-up practice and anthropogenic climate change. These starting points were consistent with others' research agendas in the Department of Geography so that when The Leverhulme Trust invited bids to initiate a major research programme on climate change in 2013, we threw our hat into the ring. A team of eight – Kye Askins, Mike Barke, Andrew Collins, Richard

Kotter and Jon Swords joined us – produced a proposal that emphasised creating new knowledge on human reactions to climate change. We were delighted to be shortlisted by the Trust, but our interview with them did not go well – our approach to 'bottom-up' thinking was not what they were looking for – and we failed to get funded.

After the initial disappointment, our reaction has been to take the positives thrown up by this experience. It was an achievement to get short-listed in the first place, indicating that we were on to something worthwhile. Thus we thank The Leverhulme Trust for setting us on our way. And, of course, there is the input of Kye, Mike, Andrew, Richard and Jon, who also have our gratitude; they helped produce the grounding on which this book is based. But the book is obviously not what a big research project would have produced. The ambitious large-scale field research that we planned was not to be. It had been scheduled to encompass both the Global North (Newcastle) and Global South (Maputo); this was now out of the question – the three of us were simply holed up on campus in Newcastle. Limited to desk-based research, the book evolved out of some intense discussions on the changing nature and content of social science contributions to the threat of anthropogenic climate change. These led to two articles published in *ACME: An International Journal for Critical Geographies*. Once again we were encouraged; being successfully peer-reviewed for publication suggested some of our out-of-the-box ideas might just have legs. These articles led directly to us producing this book; they are built on as two of our chapters. Hence we thank the *ACME* editors for their open-mindedness and support.

Open-mindedness is all we ask of our readers. Our purpose is to contribute to understanding the 'anthropo' part of studies into anthropogenic climate change. This is formally the task of social science – as human geographers we come under this label – but the research that dominates both theory and practice of 'climate science' is that of physical scientists. Thus, inevitably, the 'anthropo' bit is relatively neglected. But in our account we go one step further: what social research there is on the 'anthropo' bit is misdirected, and the mainstream social science approach is deeply flawed. To explore this situation we turn to the celebrated urbanist Jane Jacobs who often showed her frustration with the orthodox thinking of the social sciences; given the stakes at play in the current climate emergency we have her profound dissatisfaction to a state of trepidation. 'Rethinking' is an over-used term in academic parlance, but this has become an urgent necessity, and this book is our modest contribution to that end. In fact, following the world-systems analyst Immanuel Wallerstein, we will call what we do 'unthinking social science', once again upping both meaning and rhetoric to fit the high stakes.

It follows that this is a strange book. Our business is not to add to the stock of knowledge, but to change the direction of knowledge creation. Thus we bring no new evidence to our argument; instead, we reassemble existing knowledge in new ways to make our particular contribution. Chapter 1 provides eight 'preparative conclusions' as 'Declarations'. These are brief statements that set the scene for our unthinking. They refute current theory and practice and point towards the need for thinking differently. It is here that we select Jane Jacobs to guide our unthinking, and Chapter 2, 'Alternate', outlines her legacy that is available for us to mine. We take from her the notion that cities are so demanding that they shape non-urban worlds to meet their ever-growing needs. In Chapter 3, 'Inside Out', we present 14 antitheses to counter conventional modern thinking, thereby bringing cities to the fore in understanding macro social change, including anthropogenic climate change. We favour radical change formulated as 'inside out' rather than the traditional 'turning the world upside down' because historical experience of the latter has merely resulted in alternative elite rulers. This argument enables us, in Chapter 4, 'Reset', to construct a fresh mindset on climate change through a new trans-modern narrative placing urban demand centre stage. In Chapter 5, 'Action', we conclude by asking the ultimate question of how to combat modernity's transformation of urban demand into what is becoming terminal consumption. The crucial importance of the sheer scale and speed of recent Chinese urbanisation is recognised; there must be an essential role for Chinese cities in finding an answer. From Declarations to Action, this book is an exercise in unthinking our modern being; we call for much more unthinking from our intended readerships: concerned scientists, politicians and citizens.

However successful or unsuccessful we have been in promising to produce an unusual book – that is for others to decide – at the end of the day it is just a book. We are who we are, and for fellow scientists our arguments are fully developed through references to relevant work by our scientific peers, with the inevitable long reference list at the end of the text. Non-scientists should not be put off by this parade of past writings. We have written the text in a more accessible style than is usual for a scientific study: more like a political tract than an academic text, it is very direct and purposive, often hard-hitting, always readable. Furthermore, our key audience is neither scientists nor politicians, but citizens worried by climate change and confused as to why doing something definitively about it continues to be beyond our grasp. More and more citizens are joining the ranks of the worried with whom we are keen to engage. To aid this process of mobilising new people we have added an Appendix containing a brief 'Primer' describing climate change policy up to the present. We consider such policy developments to be largely necessary but wholly insufficient. Some readers might want to start with the Primer as a reminder of the issues, before commencing with our Declarations. Beyond this simple aid, the task of marshalling 'inside out' thinking for reimagining climate change, notably within the contemporary world of the global electronic communication explosion, is beyond the skill-set of the authors. Thus one purpose of this book is to plead to others – scientists, politicians or citizens – to take up this urgent task. This follows from our main purpose to encourage unthinking our world. This is what our book does. By its very nature there will be much that many readers will find difficult to agree with. Okay, as long as you provide more useful unthinkings – that will be better for everyone, especially our grandchildren.

Finally in this Preface we need to explain the dedication of this book to grandchildren across the world. This might appear somewhat pretentious, but it is genuine; the book is written for them in the sense that it is intended to intervene on their behalf. The developmental working title of The Leverhulme Trust project that led to this book was 'Don't fry the grandchildren'. We were not sure which generation's grandchildren were in peril, but catastrophic global risk was certainly on the horizon. In fact, before any actual 'frying' there will likely be many social traumas resulting from the ecological decline of the Earth as the home of humanity. Living through a rapidly changing worldwide climate crisis does not bear contemplating. But we must. And it's urgent. Here is the countdown for responding to the recent United Nations Intergovernmental Panel on Climate Change's assessment about when we might need to have made unprecedented changes in our behaviour in order to avoid precipitating environmental catastrophe across the world:

~~2018 2019~~ 2020 2021 2022 2023 2024 2025 2026 2027 2028 2029
2030?

1

Declarations: Root and Branch Unthinking

The enormity of the problem has only just dawned on quite a lot of people.... Unless we sort ourselves out in the next decade or so we are dooming our children and our grandchildren to an appalling future.... What we do now, and in the next few years, will profoundly affect the next few thousand years. (David Attenborough, speech at Davos, Switzerland, January 2019)

Introduction: unwelcome statements of the evident

Things are not going well in the world of climate change policy and action. There are myriad initiatives and innovations, not least in alternative energy, but somehow it never seems enough. The threat of global warming and its numerous dire consequences are unrelenting. Why? It is not that we do not know what is happening; the science is conclusive, extensively reported and widely accepted, albeit not universally. Obviously humanity's overall responses to these authoritative warnings are wholly inadequate. But this answer only elicits another question, why inadequate? It is this second 'why' that we attempt to answer in this book. Put simply, we think that the basic framing of the global climate change debate – scientists providing compelling evidence so that governments can collectively produce the urgently required policies – is deeply flawed.

We begin this book with a set of *Declarations*. These are eight statements of our starting position, blunt pronouncements that are the foundations of our thinking. Hence we do not discuss and debate current climate change policy and action; we treat its failing as a given. We do not offer ways of improving what is being done through building on successes and avoiding known pitfalls. This is not a 'how-to-do-better' manual. We do not belittle the very important work being done in the field, but these efforts are known to be wanting overall. They don't add up to anything that looks like an adequate solution. Despite multiple exhortations to sort it out there is a feeling of exasperation every time a new set of climate science warnings is published. It seems to be not so much a matter of 'what can be done?' as admitting 'what can't be done?' We present what we think are intractable obstacles that appear to be blocking a credible path ahead and suggest an alternative way forward.

The first group of Declarations identifies fundamental difficulties: we call this 'clearing the decks'. These are stark statements that underpin all our subsequent discussion through default – treated as a given, they are simply left as such. No more discussion. These assertions are our starting point and we ask the reader to bear with us. Between them they generate an impasse in how to think about climate change policy and action. The second group of Declarations suggest a way out through critical 'unthinking'. We derive this idea from Immanuel Wallerstein (1991), where he calls on social scientists to 'unthink' their continued use of 19th-century paradigms, and have added some 20th-century paradigms. Specifically we indicate why we have chosen Jane Jacobs, the prominent urbanist, as our guide for our unthinking, while acknowledging that others of her critical ilk provide for alternative unthinkings. The point is that we urgently need to use ideas from unusual contrarian figures to shine new light on failings of conventional thinking.

The Declarations are ordered to indicate the overall rationale behind the book: early Declarations provide the reason for the book, later Declarations outline the purpose of the book, and in between there is a progression from the general to the more particular, as embodied in Jane Jacobs' oeuvre. In their different ways each Declaration is highly contestable; collectively they may appear to put us beyond the modern pale. That is where we need to be.

Clearing the decks

We come from a rarefied world called academia. Between us we have many decades of experience in higher education, mainly in the UK, but also in several other countries within both the Global North and Global South. There are two important things we have particularly noticed about our privileged world. First, its growing size has long necessitated that its inhabitants become more and more specialised in their work – in the old adage, to succeed in this place it helps to know more and more about less and less. Second, and despite the efforts of some inhabitants, academia remains somehow separate, even seen as 'above', the common concerns of 'ordinary' people. Thus to call something 'academic' actually provides a connotation of irrelevance. The point we want to make is that these two characteristics are preventing this important sector of work contributing at the best of its abilities to what is the most important issue of the 21st century: anthropogenic climate change.

Within academia we work in a section called 'social science', and as social scientists studying anthropogenic climate change we are trying to avoid the two negative features of our working world as previously discussed. This is a key purpose of the Declarations – to show where we are coming from. To this end they are simple, transparent and succinct. None of the Declarations are original – we are reworking some well-rehearsed ideas – but the manner of their merging is intended to provide a fresh thought-provoking agenda. Agree or disagree, at the very least they are intended to introduce what urgently needs to be addressed by all of us, as producers of commodities and ideas, and as consumers of commodities and ideas.

Declaration 1. The kind of problem anthropogenic climate change is [1]: A complexity duet

When confronted with a problem the first step is to decide the kind of problem we are dealing with. In policy terms there are two main categories, simple and complex. Simple problems involve just a few, usually just two, key variables that are related in a straightforward manner – for instance when one increases, the other declines. This allows the policy-maker to identify a 'trigger' that can be operated to manipulate a policy target in a preferred direction. Governments have devised myriad policies in this manner to good effect. A classic example occurred in 1950s Britain where 'smog' – a combination of smoke and fog causing many deaths – was dealt with through the Clean Air Act 1956 banning the heating of houses through coal fires. It worked; no more smog. On a global level the hole in the ozone zone over the Antarctic discovered in the 1980s was dealt with through the United Nations (UN) by banning the use of chlorofluorocarbon compounds (CFCs) in manufacturing. It worked; the hole is reducing. But this approach will not work when the problem is complex.

Anthropogenic climate change is not a simple problem. This may seem an extremely obvious statement, but it needs heeding in policy-making where triggers and targets dominate agendas. If carbon emissions are the prime cause of climate change over the last 200 years, then agreement to curtail use of carbon-based energy should surely do the trick. Supposedly job done (as long as all keep to their promises); global climate should begin to stabilise. NO. The achievement of the Paris Agreement in 2015 is certainly not worthless, but how worthwhile is it? The focus on carbon reduction does not give due weight to why carbons are being consumed in the first place – it tackles supply without full cognisance of demand. And bringing the latter into the argument throws the complexity spanner into the works.

Complexity is accepted in climate change science in which modelling involves many variables changing simultaneously; global climate is treated as an organised whole within which all variables are interconnected. But this is not the only organised complexity in operation; humanity is organised globally with myriad economic flows interlaced with cultural differences and variegated political processes all moving simultaneously, and this interconnected world is a highly complex social whole. Anthropogenic climate change is therefore doubly complex, the result of critical interactions between these two organised complex wholes. This complexity duet involves both climate and society working through their separate operating logics while at the same time intersecting to disrupt and alter each other's complex unfolding. The resulting developments are anything but simple, and should not be treated as such.

Anthropogenic climate change is a problem deriving from incessant crisscrossing of two organised complexities. Policies that do not appreciate this are doomed to fail.

Declaration 2. The kind of problem anthropogenic climate change is [2]: Existential threat? Existential!!!

Having identified anthropogenic climate change as a complex problem it must be added that from humanity's perspective it is unique. It is qualitatively different from all other organised complexities scientists study and make policies on because for humanity it presents an existential threat. Take a moment to think about this. It is simply incredible that we should be writing about such a threat. The adjective, 'existential', means the reader's grandchildren, great-grandchildren or great-great-grandchildren literally will not exist. Just having to state this chilling possibility is shocking beyond belief.

We are not the first people to face an existential threat. Historically all the great conquerors became famous or infamous for their ability to wipe out cities and civilisations. Their potential victims had to devise policies of survival, and there were three basic options: resist, accommodate or flee. However traumatic, in policy terms, these were simple answers to an unambiguous problem. In the second half of the 20th century the nuclear war threat was viewed as existential during the Cold War. But again this was essentially a simple problem and there was sufficient rationality on both sides to negotiate out of the confrontation with first, a test ban treaty, and then, a nuclear proliferation ban. These policies were not perfect but the existential threat was largely contained. Since 1992, the United Nations Framework Convention on Climate Change (UNFCCC) has attempted to treat existential threat in a similar way. Although the many negotiations themselves can be considered complex proceedings, the whole purpose is to bring the parties together in a practical policy; in other words, it is a process that has simplification at its heart. As noted above, it has created a carbon reduction policy as a very partial treatment of the complexity duet that is anthropogenic climate change.

In the last 500 million years five mass extinctions have been completed whereby a large proportion of species disappeared in a relatively short time frame. We are now living through the sixth such mass extinction, which is down to the way humans have treated the environment in their growth, spread and development. Climate change's existential threat therefore represents the discomforting irony of our species being a victim of our own generated mass extinction. Perhaps we deserve it – reap what you sow and all that – BUT....

We invoke a human preservation imperative. Surely the undeniable human ingenuity that got us into this trouble can get us out of it. A possibility, but only at a stretch, and what needs stretching first is our thinking to ensure it encompasses the complexity duet.

Declaration 3. Not fit for purpose [1]: International relations as the worst possible route to a solution

The Paris Agreement on reducing carbon emissions has been criticised previously – we might put it into the category of 'necessary, but not sufficient'. But this would not be quite right. It suggests that the 'sufficient' can be achieved with further policy work and negotiation at further UN Climate Change Conferences. NOT SO. Aggregating simple solutions can never add up to a satisfactory complex resolution. The UN programme is flawed precisely because it is by the UN!

Suppose you were of a malicious turn of mind and wanted to design a framework for decision-making at the global scale that was sure to fail. One key element would be to suggest multiple separate decision-making bodies that were extremely competitive. With each body primarily concerned to do what's best for itself, coming to any agreement would be extremely difficult. This is, of course, a description of our current international relations system of some 200 different countries, with governments committed to furthering their national interests and diplomatic corps protecting those interests against neighbouring countries and larger countries in general. In this competitive world, all states behave to enhance their importance; the larger countries vie for top position. And with so many different countries there is the perennial problem of some countries freeloading off others, leading to innate suspicion of each other's motives. The UN is part of a process to inject some system-wide cooperation into this politics, but it is always up against myriad positions of self-interest: this is what makes the Paris Agreement such a diplomatic achievement. But the UN cannot transcend what it is – a club of states defined by their territorial sovereignties as separate, independent decision-making bodies. We have already made the point that the Paris Agreement is the product of a political process that is predisposed to produce simple common denominator policy prescriptions. Now we can add that this is as good as it gets.

This negative prognosis is based on discussing states as a general category; if we look at the specifics, even more obstacles arise. Contemporary states vary immensely in their size and wealth, which translates into a wide range of capacities and competence. Despite UN agreements it has become clear that most members have not produced something as basic as credible carbon inventories. This is likely to be because of either capability issues – poorer states do not have the organisational means – or political issues – richer states have short-term alternative prerogatives related to governments staying in power (the electoral cycle). This means that even if the UN managed to get all states on side, it would remain doubtful whether 'global' policy could ever operate through the joint action of states.

But, you might respond, international relations are all we have for negotiating at the global scale. This would be a problem if a solution to anthropogenic climate change was indeed to come from multiple simple territorial solutions, but this has also been ruled out. We have to admit that international relations are simply not fit for purpose in this context. We have to think in a different way.

Declaration 4. Not fit for purpose [2]: The scientific work that disregards its anthropogenic component

The two parts of the complexity duet are treated very differently in the scientific input into anthropogenic climate change policy-making. While there is a large volume of physical science and social science research on both complex wholes, only the former is treated as such in policy thinking and action. Humanity is treated in a very piecemeal manner; bits of relevant ideas and research are brought to the policy table as and when necessary, but there is no notion of global process on the social side of the duet. Figure 1.1 is a stark depiction of this reality.

This bizarre situation is quite explicit if we take just a cursory look at the organisation of the Fifth Assessment Report of the main scientific body advising on climate change, the UN Intergovernmental Panel on Climate Change (IPCC), which provided the evidential foundation for the Paris Agreement. Its work is divided into three working groups: (I) The Physical Science Basics; (II) Impacts, Adaptation and Vulnerability; and (III) Mitigation of Climate Change. The Physical Science Basics is not only a repository of the latest 'hard science' findings; it is also set up to provide a coherent description of climate change in the role of instigator, leaving humanity merely as recipient (impact) or responder (mitigation). The latter roles bring forth a potpourri of important ideas, for instance on 'Coastal systems and low-lying areas' for impact and on 'Buildings' for mitigation, but without an overall social framework, there is not even a notion of a globalising whole. The closest is a chapter on 'Key economic sectors and services', but this is a descriptive listing rather than an integrated analysis, and in any case is only one chapter out of 60 in the report. The result is that at UN Climate Change Conferences the only discernable 'social science theory' seems to be 'where there's a will, there's a way'. David Attenborough expressed this sentiment succinctly at the 2019 World Economic Forum in Davos: "If people can truly understand what is at stake, I believe they will give permission for business and governments to get on with the practical solutions." This is so very naïve, totally disregarding social science.

The question arises as to why the physical complex whole is treated as a given while the human complex whole is missing. It does not help

Figure 1.1: Anthropogenic climate change without social science

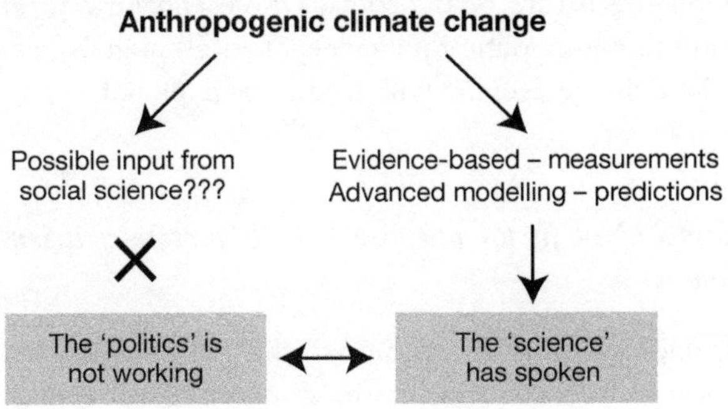

Anthropogenic climate change

Possible input from social science???
Evidence-based – measurements
Advanced modelling – predictions

✗

The 'politics' is not working ⟷ The 'science' has spoken

that the research is compiled under the auspices of the UN, with its multiple state interests, but more basic is the inherent political nature of any social science framing of contemporary humanity. As any glance at the huge literature on globalisation will confirm, the topic is a highly contested matter. This should not be surprising. Social science will always be different from physical science because social scientists are part of the subject matter they study: they peddle findings about humanity. This has many implications, not least that humanity, or parts of it, can consciously disrupt findings through people changing their behaviour in reaction to those findings. But it would be disastrous to react to this real issue by ignoring social science as unreliable 'soft science'. Although not realised by its advocates, those adopting a 'where there's a will, there's a way' approach are actually employing an elementary instrumental theory of the state that is dangerously defective. The alternative theory is to understand that the power of states to develop policy is premised on their ability to simplify complex issues, a standardisation of ideas and practices that makes implementation feasible. This basic understanding leads to critical engagement with policy-making to avoid over-simplification undermining the purpose of a proposed policy. A salutary example was the UK government's tax policy of promoting diesel cars over petrol cars in the early 20th century to reduce carbon emissions. It was subsequently found that the particulates emitted by diesel cars were actually more of a health hazard than the carbon emissions they were intended to replace. This retrograde step was the result of what can only be called a 'carbon fixation', a policy simplification resulting in a misguided 'political will'. Harking back to the previous Declaration, there is an international relations policy lesson here, of course....

Thus social science should be celebrated for its diversity as a means of dealing with complex reality. This is all to the good – a multiplicity of ideas and opinions is healthy, a mark of us being human. But its downside is clearly that it does get in the way of obtaining a position on which to build policy across the complexity duet. Can there be a meaningful consensus on the nature of the social whole and how to change it? Certainly not through political policy negotiations in global conferences – other quite different avenues will need to be explored.

Declaration 5. Not fit for purpose [3]: The state-centrism of the social sciences

The social sciences are relatively new to the pantheon of sciences, and their position as such continues to be disputed. They emerged in the

19th century located between the physical sciences and the traditional humanities through using the methods of the former on the subject matter of the latter. By the end of 19th century, with European and European settler states increasing their policy interventions in their respective societies, social sciences developed to provide evidence and theory in the new policy-making circles. In a time of great reforms, when politicians were steering different routes between progress and order, the discipline of economics was created to help reform economic policy, political science contributed to political reform, and sociology informed new social reform policy-making. This trilogy of social science disciplines claimed to cover all aspects of modern life, and in the process helped make and promote ideas and concepts of national economy, national politics and national society, each territorially coterminous defined by state sovereign boundaries. The power of this nationalisation of social knowledge should not be under-estimated: today we still take it for granted that there are about 200 distinctive economies, politics and societies because that is how many states there are.

The fallacy of this thinking was clearly revealed in the second half of the 20th century when decolonisation led to a rush of new states and the social sciences were brought in to combat poverty through development programmes: there was to be economic development (economists advising industrialisation); political development (political scientists advising democratisation); and social development (sociologists advising modernisation). The basic concept was that every country following these policies would climb a ladder of development, with the USA representing the top. New terminology reflected the optimistic beginnings: rich states became 'developed countries' (that is, implying 'job done') and poor states became 'developing countries' (that is, on the way to 'catching up'). But this was not to be. The real world of poor countries turned out to be somewhat more complicated than this simple social science devising of policy based on repeating the 19th-century experiences of rich countries. There was never a world of multiple separate ladders for each state to climb, and the rise of corporate globalisation finally put paid to this national denial of one worldwide complex whole.

Today the state-centric thinking embodied in the operation of the social sciences leaves them stranded at a limiting geographical scale. They were never developed to engage with a global social whole, but it is this scale of analysis that is necessary for understanding and dealing with anthropogenic climate change. There are two basic reasons. Traditional social science practice cannot be simply 'up-scaled' from state to global level because the latter social complex is so unlike the entity that is a single nation-state. And just like the poverty example, anthropogenic climate change will not be solved by an aggregation of separate environmental policies by lots of governments, however well intentioned and informed.

Every social institution is initially set up for a particular purpose, and over time that purpose erodes and the institution adapts to continue its relevance, but there comes a time when it is found wanting and its usefulness wanes towards an ending. State-centric social science has been found wanting by anthropogenic climate change.

From impasse to unthinking

Our Declarations have reached a dangerous impasse: there is an acute need for a critical social science input for dealing with the climate emergency. But such a social science does not credibly exist, customised to meet this end.

This is not to say there are not large numbers of social scientists doing important research on social, economic and political aspects of climate change. For instance, in the UK, the Academy of Social Sciences has published a series of 12 booklets on 'Making the Case for the Social Sciences', and 'Climate Change' features as No 3 (between No 2 'Ageing' and No 4 'Crime'). It consists of cameo descriptions of 16 research topics, presumably selected for their quality and relevance. And that is it: climate change used to promote social sciences – no real sense of its ultimate importance or urgency, just another way the social sciences can portray itself as useful to its main funders, the British government. No doubt similar marketing of the social sciences is occurring in comparable ways in other countries across the world. Put simply, the existential bit is missing!

There are numerous examples of social science reports, books and articles that address anthropogenic climate change as existential, but these do not constitute a coherent body of knowledge that is up to the challenge. Uncoordinated, often blindly disciplinary, they do not adequately deal with the 'anthropo' half in the way that physical scientists have been able to harness their knowledge through the UNFCCC, a process now a quarter of a century old.

It is our purpose to begin a discussion on how to find a necessary means of escape from this impasse.

Declaration 6. Critical unthinking [1]: The need for new social knowledge

The resulting need for a new and different social knowledge is widely understood, if not so widely trumpeted. Today the social science created in universities across the world constitutes a minority of formal social knowledge production. There is a mixture of think tanks and professional service providers commissioning and writing reports on all manner of economic, political and social issues for both public and private clients negotiating the new corporate global world. These new knowledge providers often operate as part of lobbyist campaigns, which means finding evidence to support a given position rather than coming to a conclusion through examining the evidence. This reversal in the relationship between evidence and findings is to be found across think tanks of all political orientations, albeit some more than others, including, of course, climate change denial. Overall this is primarily corporate social knowledge for a corporate world.

Creation of social knowledge in universities has also changed in response to this corporate world. There has been a large expansion in university social science but it has not primarily involved the trinity of founding disciplines; the boom has been in business schools worldwide, these new university departments with their shiny new corporate-looking buildings. Covering some aspects of political science and sociology, their main subject is economics. So how do they differ from economics departments? It is quite straightforward, actually – the latter continue to focus on understanding how national economies work, whereas business schools have a much wider brief encompassing globalisation in understanding how firms work. Originally known as 'management schools' teaching 'management and organisation science', they have become a major source of researchers for corporate social knowledge production, as previously discussed. A measure of their success is that the number of MBAs (Master's of Business Administration) graduating every year outstrips economics PhDs (the highest research degree) by multiples of thousands.

These social knowledge changes are not good for combating anthropogenic climate change; they are actually instruments to maintain and boost the very human processes creating climate change. But there is another emerging social science that is taking a very different path. For this social knowledge it is vital that we begin 'unthinking' a core 19th-century idea: the inevitability of 'progress'. Unthink, because progress is so deeply embedded in modern society that we need to consciously resist. Progress, transmuted into 'development' as intrinsically good in the 20th century, is the supreme transformative idea that has held sway, undergirded by continual technological advances, for about two centuries. Doubts were expressed in the late 20th century that ultimately the Earth was simply not big enough to sustain progress as perennial economic growth. Anthropogenic climate change is the upshot that has made this position undeniable for humanity.

The top of the erstwhile development ladder that poor countries were aiming to reach was called 'high mass consumption'. Thus the world envisaged was one with everybody consuming at USA levels. There was always a dollop of hypocrisy in the selling of this idea; now we see it for what it is – impossible. The vital unthinking has to be honest in saying that development as we know it cannot continue. Only by starting from this position can there be a social knowledge fit for purpose in providing necessary 'new social science' inputs to understanding the global complexity duet. Note that this inevitably breaks with contemporary politics, where promising a better life as more consumption is endemic; competing parties for government only differ on the particulars of how this might be achieved.

Declaration 7. Critical unthinking [2]: Jane Jacobs, turning the world inside out

Politicians will not solve the existential problem of anthropogenic climate change. This negative conclusion of our unthinking requires a positive counterpart that directly addresses meanings of development. We have chosen the work of Jane Jacobs (1916–2006) for this purpose.

Here is a classic example of Jacobs' unthinking. In her book (1970), where she grapples with the idea of development, a key chapter is entitled 'The valuable inefficiencies and impracticalities of cities'. On the face of it this is a celebration of bad urbanism, defined as cities not working well. For sure most people living in cities find these features annoying and frustrating; at any point in time city governments are trying to eliminate them, and today we have corporations promising their banishment through smart technology. But this conventional thinking misses the point. For Jacobs, a city is a continuous process of development; inefficiencies are both a sign of a city's current innovative dynamism (for example, it is outgrowing its infrastructure) as well as a stimulus for future innovative dynamism (for example, it is a challenge to create new ways of moving). But any ensuing new arrangements will, in turn, be victim to the ongoing vagaries of city development, resulting in new stimulating impracticalities. In the very unlikely event that some short-term policy of urban change did actually iron out all problems and create the perfect efficient city, then, from Jacobs' perspective, the city would lose its dynamism and fade away into history. It is not that inefficiency is good, but that it is inevitable and necessary to maintain the complex vitality that is the city.

Who is Jane Jacobs? The simple answer is a freelance writer who never held a university position. Her road to intellectual prominence started with architecture journalism leading to a book in 1961 that is still in print and famously decried 'progress' in city planning that took the form of favouring the building of freeways at the expense of destroying vibrant communities. Her later books have contributed to a wide range of social science knowledge so that it is impossible to pigeonhole her into any one discipline. This means that she most certainly cannot be labelled as 'multidisciplinary' or 'interdisciplinary' since she did not respect the disciplining of academic disciplines in the first place. Her unthinking is indisciplinary (Taylor, 2013: 2–4).

A crucial feature of Jacobs' work that we take forward is her commitment to bottom-up process and action. But with Jacobs, of course, this is not as straightforward as it initially appears. First, she was a highly successful community activist and is still revered for this role in New York and then in Toronto, and is followed by activists in cities across the world. For this she is a left-wing icon. Second, she was a great champion of entrepreneurialism, and therefore her ideas on development are centred on free markets. For this she is a right-wing icon. Her unthinking takes her beyond current political labelling. She is NOT about 'turning the world upside down' – you just get an alternative top-down politics – but rather turning the world inside out to create really new and different social relations.

Declaration 8. Critical unthinking [3]: Alternate invitation

Jane Jacobs never addressed the question of anthropogenic climate change; she provides us with no answers. But she is important to developing the 'anthropo knowledge bit' for two reasons. First, as previously argued, she represents the kind of thinking we need. She is one among numerous unconventional writers who can contribute to how we unthink considerations of our current existential predicament. We have chosen to feature her in particular because of familiarity with her work due to our own research. But we will need to collect vital unthinkings from many sources in order to develop a suite of ways to transcend 'progress' as the preordained future for humanity. To this end the need is for multiple alternate probings because we are not interested in simple alternatives (clever utopias): it is exploration of a bottom-up process in a complex whole that is required. Alternate implies an ongoing mechanism generating qualitative changes back and forth (like a dynamo). This is in keeping with the kind of problem we are dealing with. We issue an alternate invitation because we need some means of collecting unthinkings on bottom-up processes far beyond our ken.

Second, Jacobs provides clues about how unthinking can link to action – what do we have to do? These Declarations have focused on ways of thinking because this is what we do as social scientists, but it is also the right place to start moving forward rationally. In conventional social science new thinking leads to policy change, but the situation we have outlined is far beyond the scope of such a linear sequence. An existential methodology requires intense partnerships between variegated researchers and diverse populations – an interaction of continuous knowledge transfers both ways. Jacobs was a great urbanist, and deriving probings from her brings cities to the fore, not in the form of 'mayors ruling the world' (God help us!), but rather as the epitome of social complexity. And this complexity is the result of development ecologies, new ways of doing and making things – culturally, politically and socially as well as economically. This has been predicated on satisfaction of the demands of multitudes of urban dwellers over many millennia. It is this urban demand mechanism that is currently manifest as mass consumption, the immensely complex core process in today's rapid anthropogenic climate change. From this perspective it is this mass behaviour that has to be tackled head on both to understand and to limit climate change. From Jacobs' ecological definition of development, we can extract a social change mechanism that transcends economic consumerism. Broadly this 'social development' maintains the dynamism that makes cities, and with them the unique creative abilities of humans that we can continue to celebrate looking forward. But the task is incredibly huge: we are contemplating changing the behaviours, and aspiring behaviours, of billions of people, a new bottom-up process that essentially casts contemporary mass behaviour as terminal consumption for our species and others. Being fundamentally anti-social, contemporary consumption actions need to be cast away as something horribly shameful.

These are our current, considered ideas, but in truth we do not know how to turn complex humanity inside out. Many more probings from critical unthinking are urgently needed. But what does our chosen means of probing, Jane Jacobs' oeuvre of explicit unthinking, point towards, beyond a general focus on cities?

Upshot: our foregrounding of cities

By hitching our argument to the work of Jane Jacobs we are foregrounding cities; we build a case for replacing international relations by urban relations as the prime mechanism for understanding and combating the global climate emergency. Cities are prominent in many studies of climate change, but our bringing cities to centre stage is on a completely different level. We take our cue from one of Jacobs' lesser-known works that highlights the innate economic power of cities to mould their environs, near and far, to meet a range of urban needs (Jacobs, 1984; Taylor, 2013). This is urban demand, the prime force behind social and economic development, creator of civilisations including our modern one, and all the time remaking nature. As such it is revealed as the crucial social process that has been profoundly entwined with environmental processes culminating in our current climate emergency.

Today we face a runaway urban demand: 20th-century American consumerism diffused across the world, which is now being augmented by China's booming urban growth in the 21st century. Today this 'consumer modernity' is vigorously reproducing itself through the Advertising-Big Data-Social Media complex. This is the amalgam of services and industries that are stoking global consumerism on and on, seemingly, until it consumes the Earth as the home of humanity. This is what we call terminal consumption.

But there is a proviso that needs to be made immediately about the apparent commonplace nature of the urban demand concept. The everyday image is of busy shopping malls, but it is so much more than this single activity. Urban demand is a complex of forces encompassing many hidden depths that are commonly overlooked. In fact, this power of cities is habitually hidden in plain sight because of our pervasive association of power with the activities of modern states. Take, for example, the supply of clean water to cities. Traditionally a local matter using rivers and wells, when these sources are outstripped by urban growth the demand for water generates engineering solutions bringing in water from afar. Thus for Manchester in the late 19th century this involved flooding a valley 160 miles away in the Lake District, linked to the city by an aqueduct. For Los Angeles in the early 20th century it took its water from a valley in the Eastern Sierra Nevada, using a 233-mile aqueduct. Both examples faced much opposition from people in the remote valleys, but their resistance was to no avail. Through legal means involving various state instruments, urban demand prevailed. The involvement of the state, in the form of an Act of Parliament and the US President respectively, should not obscure the fact that these were simply political mechanisms for satisfying an all-powerful urban demand: if people are to live and work in cities, they have to be supplied with water, 24/7. There was never any doubt that these cities would get their water; the only choice was who would get flooded. This singular manifestation of the understated power of cities is multiplied over and over again with ever-changing consumption patterns. In today's society where most people live in cities, generalised social needs are represented by a new urban demand complex: energy demand satisfied 24/7 by wires and pipes from afar; radio and TV demand satisfied 24/7 by transmitting signals from afar; internet demand satisfied by satellites from very afar, and so on. Modern living is urban, and is so very demanding – and increasingly terminal. And all these citizens still have to eat....

Thus our take on the climate emergency culminates in the question, how can we counter terminal consumption? This simple question begets an immensely complex and difficult exercise in social transition drawing on our critical unthinking. This book charts a route to understanding cities demanding the Earth. In the last chapter we address a fundamental need to reinvent the city, to generate a new urban demand complex that is in sync with the global environment.

2

Alternate: Jane Jacobs' Legacy

Introduction: Jane Jacobs beyond urban celebrity

Zipp and Storring (2016b: xvii–xviii) have made a powerful case for revisiting Jane Jacobs' rich intellectual legacy:

> There's no doubt ... that this is just the right time for "more Jane Jacobs" ... to reimagine Jacobs herself as more than a symbol of urban sorrow or urban triumph. Always idiosyncratic and unorthodox, often to risk being wrong if it means reorienting stale conventional wisdom, she pushes beyond the familiar alarms to see urban transformation as a source of radical possibility and opportunity.... Jacobs was perhaps our greatest theorist of the city not as a modern machine for living but as a human system, geared for solving its own problems.

The description of her as 'greatest theorist of the city' is probably derived from a Planetizen survey (Planetizen is a US urban planning news website) in 2009, where Jacobs was ranked first among the 'Top 100 Urban Thinkers' by a readers' poll (Goldsmith and Lynne, 2010: xxiv). A similar survey from the same source in 2017 again ranked her first, this time as 'The Most Influential Urbanist' of all time (see www.planetizen.com/features/95189-100-most-influential-urbanists). The accompanying citation refers to two aspects of her work: the book *The Death and Life of Great American Cities* (Jacobs, 1961) and her battle with Robert Moses over New York planning issues in the 1950s and 1960s. This surely represents her prime legacy (Hirt, 2012a). Still in print today and always mentioned on the front cover of all her subsequent books,

Death and Life catapulted Jacobs into literary stardom as an authority of things urban: the book is credited with igniting paradigmatic change in the theory and practice of city planning.

But Jacobs' reputation rests on much more than her writing. She was a protestor and activist, defender of communities and advocate of civil disobedience. Her successful struggles to stop Moses physically destroying communities complemented her book and have become the stuff of legends, a modern David and Goliath story (Flint, 2009; Lang and Wunsch, 2009), even making its way onto the silver screen – see Altimeter Films' *Citizen Jane: Battle for the City*. Laurence (2016) has shown that it was not quite so simple – Jacobs was a seasoned professional writer before she wrote *Death and Life*. However, as Zipp and Storring also point out, there is more to Jacobs than this urban celebrity. Post-1960s there is a steady stream of new Jacobs' books that are, in their different ways, equally imaginative and original. Thus the extensive legacy of Jacobs far outstrips her halting of Moses' bulldozers. As well as city planning, she makes radical incursions into other areas of thought and practice in economics, philosophy, politics and environmental and education studies. Although typically dwarfed by *Death and Life*, these later challenges to orthodox thinking are of equal relevance in deriving the legacy to be discussed later.

We have constructed a specific Jane Jacobs legacy to fit our particular needs as set out in the Declarations in Chapter 1. Jacobs makes a few references to climate change, but there is no systematic pursuit of the subject in her work. For instance, we might expect her final book *Dark Age Ahead* (Jacobs, 2004) to tackle this matter, but all we have is a short reference to the Premier of Ontario being against signing the Kyoto Protocol because it would put Ontario jobs at risk in competition with the USA (Jacobs, 2004: 60), which simply endorses our Declaration 3. Elsewhere there are brief references to fossil fuels generating global warming where she defends economic development as preferable to stagnation: only through development will the alternative energy sources be created to replace polluting fuels (Jacobs, 2000: 129; Jacobs, 2002/2016: 396–7). So in this book we are not taking forward a copious body of ideas she has developed on this topic. Rather, we are writing in the same spirit that van den Berg (2018) employs: in her gender studies she uses Jacobs in order to better understand 'genderfying' the city. Although Jacobs did not write as a feminist – she did briefly comment on gender issues (Jacobs, 1994b/2016: 328–36; see also Berman, 1988: 322) – van den Berg does show that her ideas, and particularly the way she worked, can be engaged with so as to contribute to a feminist discourse. This is how we use Jacobs discursively in this book. The Jacobs legacy we outline is built to inform broadly how we might think about and engage with anthropogenic climate change.

We have selected five areas of Jacobs' thought and practice that we believe can have direct relevance to confronting the existential threat that is climate change. These are knowledge, economics, history, politics and Nature. Cities appear to be conspicuous by their absence in this list, but in fact they are a unifying theme throughout this sequence of critical topics. We conclude with two specific commentaries on this oeuvre. First, we interrogate the concept of urban demand, the part of Jacobs' ideas that has most relevance to our subsequent discussions. Second, because Jacobs is often treated in relative isolation as an 'urban celebrity', we need to link her ideas to other scholars' radical thinking that we will subsequently use. From the perspective we are developing, these 'fellow travellers' both validate and augment our Jacobs' foundations.

Knowledge building: complexity and curiosity

Above all, Jane Jacobs was a knowledge builder. Laurence (2016) describes the context in which she carried out this pursuit (see also Hirt, 2012b; Szurmak and Desrochers, 2017). In this quest she maintained two guiding principles: a respect for complexity, and curiosity as a necessity. This 'how and why combo' enabled her to create new evidence-based theory across a range of fields of knowledge.

The starting point for considering Jacobs' contribution is the oft-quoted final chapter of *Death and Life* (Jacobs, 1961) entitled 'The kind of problem a city is'. Musing on previous chapters describing tragedies of city planning theory and practice, Jacobs concludes that it all comes down to a fatal misunderstanding of the nature of cities. She draws on a recent interpretation of the history of scientific thought by Warren Weaver (1948). He identified three different types of research problem, each to be solved by a different methodology. First, problems of simplicity – cause and effect processes involving few variables – are soluble through basic experiment. Second, problems of disorganised complexity – the interrelations between large numbers of variables – are soluble using statistical probabilities. Third, problems of organised complexity – simultaneous interrelations of many variables organised into a functional whole – are soluble by holistic understanding. It is this latter case that we confront in cities, and therefore policies based on cause and effect or statistical predictions are doomed to failure. It is not just that they are partial, but by not comprehending this particular complexity, resulting policies inevitably lead to multiple unintended consequences, many harmful to cities. This view of science remains vital to Jacobs' thinking, and she continued to use it many decades later (Jacobs, 2004b/2016: 441).

Jacobs (1961: 454) draws three methodological lessons, which she calls 'important habits of thought': to focus on processes; to start inductively; and to privilege small 'unaverage' clues. This last is the most unusual instruction; it replaces statistical generalisation with curiosity – why such an odd occurrence? This puzzle solving has also been a key source of criticism of her methods. Her research draws on inquisitive observation and veracious reading to provide anecdotes that pepper her writings. However, these are cited as evidence of a deficiency in the 'normal standards of scholarship' (Harris, 2011: 80; see also Duranton, 2017). Cichello (1989: 123) describes more broadly the situation as follows:

> There is, clearly, a common core of criticisms, a familiar litany of complaints: she lacks rigor and careful observation; her presentation of data is sporadic and selectively chosen to make her case; her examples and categorizations are incorrect or inconclusive. Her time perspective is distorted, and her theory in general is unsupported by models or statistics.

To be sure her method of investigation is 'messy, muddly work' (Keeley, 1989: 35; see also Szurmak and Desrochers, 2017: 23), a journey of trial and error (Cichello, 1989: 132), the exact opposite to the hypothetical-deductive method favoured by economists and many other social scientists. But this is actually how the ways of good research operate; the neat, logical sequence that is presented in scientific papers is just that, a presentation, and one that tends to conceal research curiosity. She accepted, however, that scientific thinking could operate in both ways; it was simply a matter of where the hypothesis was formulated in the research process, at the very beginning or near the end (Jacobs, 2004: 68–9). But whatever the method, curiosity remained the key. Keeley (1989: 33) calls this her 'way of wonder': expect the unexpected because all research requires those Eureka moments (Jacobs, 1993/2016: 319).

In two of her later books, *Systems of Survival* (1992) and *The Nature of Economies* (2000), Jacobs experiments with a very different form of presentation to relay her puzzle solving. Didactic dialogue is used by invented characters who argue and debate alternative positions in a fictional novel format. In this way Jacobs presents her argument as an unfolding narrative of disputes tested through evidence and through which she hopes her readers will get involved. Both books provide further knowledge building. *The Nature of Economies* will feature strongly in later sections, but here we focus on *Systems of Survival* that investigates alternative moral codes in the world of work.

Jacobs explores the moral contradictions in doing the 'right thing' in a working environment. Two behavioural traits that are generally seen as being good are loyalty and honesty, but these can often be in conflict. This is obvious in job recruitment – should a new management job be kept in house or go to the best-qualified candidate? From such simple conundrums Jacobs builds two separate moral syndromes – the practical knowledge that is required for success in two different types of work, which she calls Commercial and Guardian. For the former, a successful economy has honest traders, for the latter, a successful polity has loyal subjects. Both these virtues are valid and necessary but only in the right context. For instance, a loyal trader (not getting the best deal) is likely to end up bankrupt; an honest general (easily outsmarted by an adversary) is likely to lose the battle. This practical knowledge has been generated through experience over millennia so that the Commercial moral syndrome validates cosmopolitanism, enterprise, initiative and thriftiness, and the Guardian moral syndrome encourages a territorial focus, obedience, largesse and ostentation. Jacobs' own knowledge building in this work is an addition to the practical philosophical thought tradition that has focused on providing advice to the ruler (Guardians of the state) while largely ignoring ethics in the marketplace (Commerce in the city) (Jacobs, 1989b: 266, 1993/2016: 295).

These moral syndromes impinge directly on the work of knowledge building beyond philosophy. Jacobs (1992: xii) points out that 'science flourishes only in societies that have achieved commercial vitality, but art can flourish in societies that lack commerce as well' (see also Jacobs 1989a: 219, 221). However, she sees academia as largely Guardian in orientation, deriving from traditional universities, but extending into modern universities as academic defence of intellectual territories: 'There are whole fields of learning that are jealously guarded as territory within certain departments of academia' (1989a: 236). Academic disciplines are not called 'disciplines' for nothing! There is an irony here in terms of marketing her books: the publisher has to indicate on the cover where each book is located in the roster of academic disciplines. Here are the results for her main books: *Death and Life* is 'Sociology'; *The Economy of Cities* is 'Economics & Sociology'; *Cities and Wealth* is 'Economics'; *Systems of Survival* is 'Sociology/Public Policy'; *The Nature of Economics* is 'Business/Economics'; and *Dark Age Ahead* is identified as dealing with 'culture'. The latter includes education, and in this, her final book, Jacobs (2004: 45) totally admonishes today's universities: 'Credentialing, not education, has become the primary business of North American

Universities'. Accompanied by a decline of scientific thinking – the rise of the 'incurious' (2004: 87) – this failure of knowledge building is at the heart of her predicted 'dark days ahead'.

Revising economics: work in the city

In the problem of marketing Jacobs' books by disciplines mentioned earlier, the two social sciences most featured are sociology and economics. As Zipp and Storring (2016a: 311) point out, this is somewhat ironic for the former because of Jacobs' frustrations with a sociology dominated by deductive thinking, and her distain for the discipline is clear: 'Sociologists never have made a science of their subject, they just do busy work' (Jacobs, 1993/2016: 311; but see also Jacobs, 2004: 83–5), and later sociologists embracing her as an 'urban sociologist' (Hirt, 2012a: 101–35)). With regard to economics her attitude is very different. Albeit very critical, she engages with economics. In reviewing her work late in life, she even goes as far as saying that she had 'learned, finally, what I was doing – revising economics' (Jacobs, 2004a/2016: 409).

This seems a long way from the Jacobs of *Death and Life* producing a 'paradigm shift' in city planning. But as Jacobs (1992/2016: 276) tells us:

> ... learning and thinking about city streets and the trickiness of city parks launched me into an unexpected treasure hunt.... Thus one discovery led to another.... Some of the findings from the hunt fill [*Death and Life*].... Others ... have gone into four further books. Obviously [*Death and Life*] exerted an influence on me, and lured me into my subsequent life's work.

Towards the end of her life she began to bring her writings together as an economics textbook (Jacobs, 2004a/2016: 406), starting appropriately with a puzzle: why can't the rich countries of the West eliminate poverty? Her answer is that 'Important pieces of economic understanding must clearly be missing or misleading' (2004a/2016: 411); hence the need for 'a new way of understanding macroeconomic behavior' (2004a/2016: 406). The textbook was never finished – we have only 25 pages provided by Zipp and Storring (2016a: 406–31) – but we know it was to be produced as a stripped-down (sans anecdotes?) amalgam of previous books.

Jacobs' engagement with economics starts with Adam Smith with whom she disputes two key assumptions. First is Smith's argument for efficiencies due to the division of labour being instrumental to wealth creation. She calls this 'Smith's mistake':

> ... division of labour in itself creates nothing.... [It is only] a way of organising work that has already been created....
> (Jacobs, 1970: 82)

Rather, economic development is the result of new work, which changes the division of labour. Thus the latter is an outcome of the economic process, not its initiator. Froy (2018) brings this debate up to date. Second, Smith's concern for the 'wealth of nations' is deemed geographically mistaken, a 'mercantilist tautology' of looking at economics through political lenses (Jacobs, 1984: 32); Smith is a 'philosopher' with a concomitant Guardian mindset (Jacobs, 1993/2016: 296). Jacobs argues, 'nations are political and military entities' and not 'the basic, salient entities of economic life' (Jacobs, 1984: 31). Because 'most nations are composed of collections or grab bags of very different economies, rich regions and poor ones, within the same nation' (1984: 32), a one (national) policy fits all is ill-advised. The failures of such national economic policies from across the political spectrum (from various Marxisms to Keynesianism and monetarism) indicate current economic understanding to be a 'fool's paradise' (1984: 3).

Jacobs' 'new way of understanding' focuses on cities and their regions. Instead of beginning with factors of production – capital, labour and land – she brings work to centre stage, specifically the contrast between old work and new work. The former is also termed 'production work', often efficiently executed by large organisations, whereas the latter is 'development work', largely a matter of trial and error where efficiency would be an inhibitor. This distinction is important because of her definition of economic development as adding new work to old work. And this process is overwhelmingly an urban phenomenon, or more specifically, a feature of vibrant cities. It is in cities that new work is created either by innovations or, more commonly, through intermingling of innovations across cities, resulting in imitations adapted to a particular city's circumstances. Jacobs (1970) observed that the creation of new work (that is, development) occurs in bursts of urban vitality. This explosive growth of new work is the cumulative effect of import replacing, producing goods or services locally that had previously been imported (1970: 146). This is Jacobs' key economic insight and is depicted in Figure 2.1. The basic external relation of a city is work producing goods to export, that are reciprocated, in Jacobs' terms 'paid for', by imported goods of equivalent value. In this initial situation in Figure 2.1 it is old work that enables these imports. Import replacement occurs when some of the goods previously imported are produced in the city. This is new work that satisfies part of the initial consumption – Consumption A1 in Figure 2.1

Figure 2.1: Explosive city growth

1. Initial situation

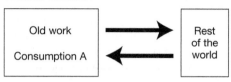

2. Import replacement and import shifting

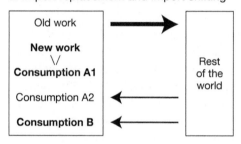

– while leaving some of the initial consumption still provided by imports (Consumption A2). This difference between the initial Consumption A and Consumption A2 is filled by new imports in lieu of those replaced, which is the import shifting to create new Consumption B. Thus the city economy has grown in size and importantly its division of labour has become more diverse. It is this combination of import replacement and import shifting that generates rapid economic development. Jacobs argues that this is 'a process of immense, even awesome, economic force' (Jacobs, 1970: 150). Evidence for this claim comes in the form of case studies of numerous cities experiencing growth spurts, as listed in Table 2.1. These many stories of dynamic economic growth events illustrate her 'anecdote' method as discussed previously; in this case, the range and weight of evidence is convincing.

For Jacobs it is this mechanism, import replacing and shifting, that defines a city as a process, a functioning economic entity. In this way cities fundamentally differ from all other human settlements – towns, villages, farms – that are economically inert in terms of development. Some cities experience just one or two economic spurts and then themselves stagnate; great cities are those that are able to generate a steady stream of such spurts. Multiple examples of import replacement provide a city with two important characteristics. First, the city has an increasingly complex division of labour, and this economic diversity provides the habitat of vibrant entrepreneurship: a case of success building on success. Second,

Table 2.1: Jacobs' explosive city growths

Cities	Explosive growth timing
Tokyo	Late 19th century
Los Angeles	Mid 20th century
London	Late 16th to early 17th century
Chicago	Mid 19th century
London	13th century
Paris	13th century
Rome	Early 4th century, BC
Scranton	Early 20th century
Chicago, San Francisco, New York	19th century
New York	Late 19th to early 20th century
Tokyo, Hong Kong, Moscow, Milan	Mid 20th century
Copenhagen	Late 19th century
Venice	10th and 11th centuries

Source: Jacobs (1970); sequence of mentions in Chapter 5, 'Explosive city growth' (pp 145–79)

the city becomes increasingly self-sufficient, producing more and more of its own consumption, thereby requiring proportionally less imports.

In contrast, where import replacing declines, there is a retrenchment into increasing old work leading the city towards economic specialisation. The resulting efficiency of large-scale old work is debilitating for entrepreneurship because it leaves fewer opportunities for diverse innovation. Detroit is the classic modern example of eventual stagnation due to economic specialisation. Jacobs (1970) illustrates this process through the contrasting fortunes of England's two great industrial cities in the late 19th and into the 20th century: the Birmingham city-region with its multiple small workshops overtakes the Manchester city-region with its large factories. Thus Detroit as 'Motown' was following Manchester as 'Cottonopolis' – it is never good to be identified by just one industrial sector.

The geography of Jacobs' development model is quite straightforward: it combines agglomerations of entrepreneurs within a city – where innovation and import replacing is done – with economic contact with entrepreneurs in other cities, through which diffusion of innovations and import shifting are enabled. In this formulation inter-city relations are always as much cooperative as competitive. As Jacobs (1970) tells it, cities need each other, and always come in groups (networks). Thus the rise of cities in new regions is often the result of beneficial relations with an existing vibrant city; every great city has to start somehow, and Jacobs

(1970: 170–9) provides numerous examples such as Venice building on its early links to Constantinople and New York on its early links to London. But once established in a region, vibrant inter-city relations continue to be necessary for further city economic success. Jacobs provides two examples: Canada's five 'hub cities (Vancouver, Calgary, Winnipeg, Toronto and Montreal) require each other: 'a hub city needs strong and many-faceted trading, information and other relationships with other hub cities' (Jacobs, 2001a/2016: 366).

But perhaps more intriguingly, Jacobs (1984: 135–55) devotes a whole chapter to 'Why backward cities need one another'. In this argument she highlights the dangers of cities in poor regions trading with richer cities because they can become mere transit places for goods to the more developed cities. To avoid this dependency relation Jacobs suggests initially promoting trade between less developed cities because they can better imitate each other's innovations. More generally, economic dependency afflicts large regions across the world, reflecting the economic power of rich cities. In the last century and a half it has been the economic demand in rich cities that has shaped a worldwide economic geography of commodity and labour flows, creating gross international inequalities. Jacobs describes the resulting multiple 'supply regions' (specialising in raw materials and exotic foods) as 'stunted and bizarre economies in distant regions', or more simply, 'economic grotesques' (1984: 59). This is, of course, the complete opposite reasoning to classical Ricardian trade theory of comparative advantage, which posits equal benefits to both sides in free trading.

So how have economists reacted to this critical engagement? Although initially ignored, Jacobs' ideas came to the fore, but mainly in one limited area: local economic growth theory (Nowlan, 1997; Desrochers and Hospers, 2007). Froy (2018) suggests a wider engagement. Jacobs' work was introduced into economics by Lucas (1988) and Romer (1986) as an externality, or non-market benefit, resulting from agglomeration. These are now widely referred to as 'Jacobs' externalities'. She is accepted as an inspirational figure in both the 'new economic geography' (Krugman, 1995) and urban economics (Glaeser, 2011). In this way her ideas have been harnessed in both radical and conservative economic thinking: respectively, Krugman (1995: 5) refers to her as 'something of a patron saint of the new growth theory', while Glaeser (2011: 11) accords 'the wisdom of the great urbanist Jane Jacobs' as a key source of his urban writings. Overall these are no mean endorsements; in fact, Lucas and Krugman are both Nobel Laureates in Economics. So where does Jacobs fit into contemporary economic thinking? Although Jacobs never cites the late 19th-century English economist Alfred Marshall, his more practical and empirical focus on economic areas (industrial districts), rather than the policy and theoretical focus on political areas (nation-states) by John Maynard Keynes (Belussi and Caldari, 2009), can now be seen as the economics tradition within which Jacobs can be reasonably said to belong.

Finally, as can be seen through her economics, Jacobs brings the process of cities to the centre of her work. Harris (2011: 65) identifies this as an 'obsession', resulting in her overplaying the importance of cities: for Jacobs (1970) cities are 'not just *a* force but in many ways *the* force' (1970: 77–8; original emphasis). So we might reasonably consider Harris to be correct; but his is an indictment from within conventional thinking; contrariwise for us it is a necessary release from state-centric thinking as required by our Declaration 5.

Glimpses of a history narrative: the Plantation Age

Jacobs was an avid user of history. We have seen how she used anecdotes, many of them historical, to illustrate processes, and how she has been criticised for this. Harris (2011: 65) likens Jacobs to a magpie collecting 'flashy items' in order to get 'her reader's attention'. And it is certainly true that she offers us a wide range of short historical stories throughout her texts. Table 2.1 illustrated this for cases of dynamic urban growth in *Economy of Cities*, and this is from only one chapter. We can add to these from elsewhere in the book in order of appearance: mid-19th-century San Francisco, an embryonic city in Neolithic Anatolia, cities of the Mesopotamian civilisation, 19th-century British cities, 20th-century Detroit, cities of the Indus civilisation, medieval European cities, early 20th-century Tokyo, Shakespeare's London, 9th-century Venice, classical Rome, and pre-classical Etruscan cities. There is no overall narrative; each makes a point in an unfolding argument. Henri Pirenne's (1925/1969) account of the rise of early medieval commerce in Europe is a rare example of influence on Jacobs' writing from her part-time college days (Zipp and Storring, 2016a: 5; Laurence, 2011: 19), and can be expanded to a wider narrative of Venice rather than Rome as the wellspring of modern Europe (Taylor, 2013: 215–28), but Jacobs does not use him this way: each historical story is an economic puzzle to be explained.

Jacobs' distain for historical narratives may be related to the way in which mega-concepts like 'modern' and 'capitalism' had distorted, in her view, how we understand the present. In the former, the narrative of being 'modern' had been captured by her New York adversary Moses to mean progress through destroying communities; in the latter, the narrative of superseding 'capitalism' had been captured by the Soviet Union to implement an equally malign version of progress as state economic planning. Of course these were two of Jacobs' prime bête noirs, and therefore the fact that she hardly ever uses either master concept for describing our present world is not surprising. Of course, this does not mean that she does not engage with modern thinkers: Laurence (2016: 8) views Jacobs as a critic of 'superficial modernism' and Hirt (2012b: 37) describes her as providing 'an exemplary critique' of 'high modernism'. Further, Hirt emphasises that she wanted to improve what was modern, not reject it (2012b: 38). Certainly she appreciated modern science and technology, and what it has enabled humanity to achieve (Jacobs, 2004: 64–5), stemming from the European Enlightenment (2004: 6). She rationalised her position by saying she was against 'beating Reason and Science into cities' while still appreciating a broader Reason and Science providing social improvements 'given to beneficiaries fortunate enough to receive them' (Jacobs, 2004a/2016: 420–1) But this did not translate into placing our times into a special period, either modern or capitalist, separated from earlier periods, which are thereby deemed to be wanting economically, culturally and politically. For Jacobs there is no process of 'modernisation', no 'transition from feudalism to capitalism'; rather, we are 'probably living in the last days of feudalism' (Jacobs, 2001b/2016: 378). An extraordinary statement considered from all conventional views (covering the wide spectrum from Left-leaning to Right-leaning histories), it does provide a glimpse of what a Jacobs' historical narrative might look like.

Perhaps stimulated by Jared Diamond's (1997) *Guns, Germs, and Steel*, towards the end of her life Jacobs did turn towards historical narrative: she planned an historical sequel to her last book *Dark Age Ahead*. She wanted to tell her story of how we got to the 'dark' position she portrayed, and to provide hope for a better outlook for humanity (Zipp and Storring, 2016a: 349–51, 459). Provisionally titled *A Short Biography of the Human Race*, she died before developing the project, and all we have is a fragment of its content in a speech delivered in 2004 (Jacobs, 2004b/2016). She takes up Diamond's argument that the development of agriculture 10,000 to 12,000 years ago created an uneven power struggle of agrarians against foragers that the latter won, leading to a world of empires. However, she departs from Diamond's narrative by emphasising that 'the powerhouses of agrarian supremacy were plantations' (Jacobs, 2004b/2016: 438). This was the Plantation Age (referred to as 'feudalism', as discussed previously), a production process that spread with empire:

> Practices perfected in the vineyards and grain, flax, olive and almond plantations of the old world were transferred to other

climates, as plantations for sugar, cotton, indigo, tea, coffee, cocoa, tobacco, coconuts, pineapples, rubber, opium poppies, peanuts, banana, spices, soybeans, and much else. (Jacobs, 2004b/2016: 439)

Three organisational precepts made plantations productive: (i) specialisation creating a monoculture; (ii) large scale – larger being more efficient; and (iii) planned production outcomes to avoid unexpected end results (Jacobs, 2004b/2016: 444).

Of course this political world of empires are civilisations, societies where cities prosper. Thus the rise of plantations is the supply side to the demand side that is urban growth. The raison d'être of plantations is to satisfy the needs of cities, whether for food, beverages or industrial materials. The gradual demise of this plantation practice is to be found in Europe in the last thousand years – starting with the 'sugar islands', European cities became supplied from beyond Europe. But plantation 'logic' extended deeply into industrialisation. This is where the mass production organised into large factories that created the manufacturing monoculture of Manchester in the 19th century enters her history (Jacobs, 2004b/2016: 444–5), as previously critiqued in Jacobs (1970). This process extended into the USA at the beginning of the 20th century as 'Taylorism' (scientific management), to standardise working practices across manufacturing (and lives on in hugely influential management consultancy firms today; see Jacobs, 2004b/2016: 445). Plantation thinking extended to land planning or zoning and suburban construction from the mid-20th century (shades of *Death and Life* here – 'Modern suburbs are caricatures of plantations' [2004b/2016: 448]), and continues in today's failing 'global city' skyscraper landscapes of a financial services urban monoculture (2004b/2016: 448–9).

In her final interview, Jacobs (2005/2016: 90) complains of the 'sameness' of contemporary urban landscapes, a global standardisation of production, but this is not necessarily a threat to import replacement because it is old work; US cities continue to create new work. Finally, she declares 'the Planation Age is no longer supreme' (Jacobs, 2004b/2016: 440). Very early signs of abandoning this monoculture mentality was found in 15th-century European cities where both printers and publishers with little understanding of their new readership markets had to improvise through trial and error to succeed. Very slowly moving into other areas of the economy as 'differentiated production', this is a creative process of satisfying and even generating new markets for new products. If this transition is successful we will enter an 'Age of Human Capital', where Guardian behaviours of zero-sum games between states (wars) give way to the Commercial behaviours of win-win transactions between cities (economics). Urban replaces agrarian just like the latter previously replaced forager: the optimism in this scenario is based on Jacobs' observation that 'Land can be held exclusively' (that is, territorial monopoly power) whereas 'Ingenuity cannot be' (that is, diffused network power).

One final point on this history: we know Jacobs' take on the transition from foraging to agriculture from her 'notorious' first chapter of *The Economy of Cities* where she proposed cities came before farming rather than the conventional view that agricultural surplus in Mesopotamia enabled cities to emerge. The latter supply theory of city origins is entrenched in the disciplines of ancient history and archaeology, and Jacobs (1970: 3–48) challenged it with an alternative demand theory: as small city networks emerged, existing foragers could not keep up with increased food demand, and agriculture was an urban invention to solve this problem. The latter idea was further developed by Soja (2000, 2010), and has been recently debated and disciplined (Taylor, 2012a, 2013, 2015; Smith et al, 2014; Ikeda, 2018). For Jacobs' historical narrative, this is the beginning of creative urban oases in an agrarian creative desert, but whose numbers grow throughout the Plantation Age until the 21st century, when urban dwellers have finally become a majority of humanity. This new Age of Human Capital provides hope for the future based on inherent myriad urban ingenuities.

Perplexing politics? New work versus old work

There seems to be a consensus that Jacobs' politics is hard to pin down (see, for example, Page, 2011: 4; Zipp and Storring, 2016b: xxvii–xxix). In the economics section we found her rigorously endorsed by both Paul Krugman and Edward Glaeser, whose respective economics are politically far apart. Page (2011) is particularly exercised by the house in New York from which Jacobs famously observed the centrality of the street to vibrant city life – a central theme of *Death and Life*. But that was the 1950s; fast forward to 2009, and that very same house is on the market for US$3.5 million. A fascinating anecdote, Jacobs seems to be hoisted by her own petard! This was a house she saved from Moses; Page (2011: 8–9) comments: 'Without intending to, or perhaps without being able to see clearly enough into the future, she made the neighbourhood safe for $3.5 million town houses'. The indictment is that Jacobs' celebrated community politics misses the wider political economy context in which housing markets based on citywide, or even international, capital eventually trump local interests. Not being anchored into one of the traditional political tendencies has resulted in her misjudging a commonplace urban outcome. If this indictment is correct, it would make a complete mockery of Jacobs' immense esteem as an urbanist.

Morrone (2017) provides a detailed discussion of Jacobs in debates over gentrification. In fact, she saw the dangers of neighbourhood success as 'self-destruction of diversity' even before the term 'gentrification' was coined (Zipp and Storring, 2016b: xxxvi). In *Death and Life* Jacobs is primarily concerned about 'unslumming', trying to prevent people moving out of neighbourhoods, the very opposite of gentrification. In her later life Jacobs understood gentrification initially as a positive process in moderation because it added to the diversity of a neighbourhood (Jacobs, 2000/2016: 358). The problem was that as a dominant process it had exactly the opposite effect, clearing out diversity through rising property prices. Jacobs sees the irony that gentrification then becomes a futile process because it destroys the very variety that made a neighbourhood attractive in the first place (2000/2016: 359). A fully gentrified place is as socially sterile as 20th-century residential suburbs – they are the latest urban 'plantations'. In the past such 'boring' places – rich ghettoes – eventually lost their attraction, and therefore we might surmise the US$3.5 million price tag is not the end of the story. But before neighbourhoods reach plantation state, Jacobs suggests several non-profit initiatives or 'whatever ingenuities' are available to maintain diversity (2000/2016: 360). This is what Zipp and Storring (2016b: viii) call her 'pragmatic politics'; Szurmak and Desrochers (2017: 37) also refer to her placing pragmatism over ideology. But it is still the case that Jacobs' ideas sit 'uneasily' within radical politics (Zipp and Storring, 2016b: xxvii).

Jacobs' unusual politics derives directly from her distinctive economics. Conventional political groupings relate to the Capital/Labour/Land division of economic process; instead, Jacobs focuses on the old work/new work division of that process. The former produces different variations of the familiar politics of class conflict; the latter implies a less well appreciated cross-class politics. Jacobs (1970: 248) considers class politics to be a 'secondary kind of conflict' situated within the realm of old work (production work) where 'well organised workers' share a common interest with their employers for their industry to prosper; the conflict is about dividing the spoils. It is secondary because it is conservative. This modus operandi is always threatened by new work (development work), which subverts the status quo. Thus:

> The primary economic conflict is ... between people whose interests are with already well-established economic activities, and those whose interests are with the emergence of new

economic activities. This is a conflict that can never be put to rest except by economic stagnation. (Jacobs, 1970: 249)

The result is sequences of cyclical political conflict based on 50-year Kondratieff waves of economic growth that generate cross-class political interests coming to the fore during phases of economic growth spurts. (Actually Jacobs does not directly focus on long economic cycles, but when they are presented to her she finds them 'very interesting' and relates them to innovations. Therefore what follows is a slight extrapolation of her ideas; see Jacobs, 2002/2016: 397–8.) We can argue that these long cycles generate three sets of special interests: (i) of those engaged in current development work; (ii) of those engaged in production work from within the previous cycle of development work; and (iii) of those engaged in production work from development work two or more cycles ago. The latter are commonly referred to as people economically 'left behind' in relatively stagnant cities and regions; they currently form the basis of cross-class populist politics (Taylor, 2017c). However, Jacobs' (1970: 250) key point is that class politics from interest (ii) typically dominate political practice because of its grounding in well-established organisation advantage (political parties). Politics concerning interest (i) are typically unformulated – they become established only later when they become an interest (ii). Thus political interests from crucial innovative processes in economic development are largely unrepresented in conventional politics. Now it should be clear why Jacobs 'sits uneasy' alongside conventional radical politics!

According to Zipp and Storring (2016b: xxxii), 'Jacob's vision is one of markets without capitalism'. This brings the discussion back to the Commercial ethics syndrome that is explicitly about free markets. A key point is that, despite its name, Jacobs (1992: 28) excludes work for 'commercial monopolies' from the Commercial syndrome and includes it alongside working for the state within the Guardian syndrome. Thus local neighbourhoods can be threatened by both public (city planning) and private (aggressive gentrification) practices. More generally, she advocates taking on 'powerful corporate interests and government monopolies' (Jacobs, 1994a/2016: 327). But this also means opposing state aid to poorer regions, identifying such policies as a 'transaction of decline' because they maintain stagnation at the expense of creating new work, the latter being the only way of overcoming poverty (Jacobs, 1984: 182–3). This thinking extends to her opinion on the work of The World Bank: in a discussion with Bank employees, Jacobs argues that since they can only provide investment via state agencies, and not directly to cities, 'it's questionable whether you ought to be doing anything' (Jacobs, 2002/2016: 404). But she is not anti-Guardian; although this syndrome includes 'raiders' and 'takers', it also covers necessary functions for a society and economy to work such as state monopoly of the use of force (Jacobs, 1993/2016: 307) and more generally, the aristocratic ethic of stewardship.

So what is Jacobs' politics? Her preference for small institutions, both public and private, suggests she is more anarchistic than socialist in her radicalism, but such a label is hardly credible. The best description remains Zipp and Storring's (2016b: viii) 'pragmatic politics'. For instance, her promotion of markets is not market fundamentalism of the neoliberal variety, but rather, recognition of entrepreneurs as creators of new work, development that eases poverty. Jacobs makes very practical suggestions about standard issues such as taxation: she is against value-added taxes because they favour large corporations with their in-house supply chains (no taxes) compared to small firms trading with other small firms (taxed on every transaction) (Jacobs, 1984: 225–6, 1993/2016: 316). All her pragmatic politics relate to self-developing cities and how their innovative potential can be protected and encouraged 'to creatively solve practical problems for themselves and each other' (Jacobs, 1970/2016: 219). Where this does not happen old work dominates, often with debilitating effects:

> Creative cities prevent the same natural resources from being exploited too heavily and too long. It is stagnant economies that become ruinous to the natural world. (Jacobs, 1970/2016: 220)

Jacobs coins the terms 'undone and undeveloped work' (1970/2016: 220) to describe this missing creative urban input of stagnant cities, and gives automobile highways, congestion and pollution in cities as a transportation example of undone work (that is, the multiple-cyclic old work of making cars has gone on for far too long).

Finally, one last point for those doubting Jacobs' political radicalism: beyond her pragmatism there is always an explicit critique of authoritarian modernism. James Scott (1998: 103–79) finds intriguing parallels between Jane Jacobs representing bottom-up/complexity against Robert Moses representing top-down/simplicity, and Rosa Luxemburg (bottom-up/complexity) versus Vladimir Lenin (top-down/simplicity). Hugely different political contexts, but fundamentally similar modern antinomies, creates serious food for thought.

Double Nature: fitness for survival

The one idea that pervades Jacobs' work over many decades is the notion of organised complexity as introduced previously in the section on

Knowledge. However, Jacobs has changed in the way she translates this systems thinking. Initially borrowed from the biological sciences, she was originally taking up the challenge to apply organised complexity in the social sciences (Jacobs, 1961: 446). As late as 1992 she was referring to 'two sorts of ecosystems – one created by nature, the other by human beings' (Jacobs, 1992/2016: 282). But in her *The Nature of Economies* (Jacobs, 2000), the basic premise of the book 'is that human beings exist wholly within nature as part of natural order in every respect' (2000: ix). In the otherwise comprehensive coverage of Jacobs' ideas by Zipp and Storring (2016a), this book is conspicuous by its neglect – the book is referenced only in footnotes, just six in total. But for the Jacobs' legacy being developed, here it is a vital source. To some extent it is an ecological rewriting and contextualising of *The Economy of Cities* (Jacobs, 1970), but with much less attention given to its essential complementary volume, *Cities and the Wealth of Nations* (Jacobs, 1984).

Jacobs (2000) treats development as a universal process covering both inanimate and animate processes that create 'significant qualitative change' (2000: 15). She posits this as the opposite of the commonly argued 'Thing Theory' of development: 'development is not a collection of things but rather a process that yields things' (2000: 32). She illustrates this by depicting natural ecosystems as conduits through which energy flows (2000: 46–7). In a tropical rain forest energy (sunlight) enters and is used in myriad reactions to create a dense, diversified environment. In contrast, in a desert this same energy flows through the ecosystem, swiftly leaving little evidence of its passage. And so it is with cities. Like rain forests their imports are repeatedly converted, recycled and recombined to create a complex economy. In other settlements imports come in, are used in production of exports, and leave behind essentially simple economies (a company town with just its saw mills is an obvious example). The key point in ecosystems is that the output from the conduit – what's left over once the process is complete (usually called waste) – is not the driver. Thus in a city economy it is not the exports that determine success: Jacobs' theory of economics is one of increasing returns contrasting with the conventional 'national' theory of economics that contrives diminishing returns (Jacobs, 2000: 63).

Diverse ecosystems are much more resilient than one-crop plantations, but they can still be vulnerable to collapse; they require what Jacobs (2000: 83) calls 'dynamic stability'. Dynamic systems must continually correct themselves to evade collapse. Jacobs identifies four resources and methods for doing this: (i) bifurcations – where instabilities are so dangerous that the only option is a radical change whereby an ecosystem divides into new processes; (ii) positive feedback loops – signals in the system trigger reinforcement of processes working well; (iii) negative feedback controls – maintain balances within the system, the key self-organisation of complexity; and (iv) emergency adaptations – addressing temporary instabilities that are nevertheless potentially devastating. City economies are ecosystems and therefore subject to these corrections, often having to react to misguided city government actions through their organised complexity. Thus Adam Smith's dynamic price mechanism – his famous 'invisible hand' – triggers:

> ... continual adjustments – by industry, labor, customers, landowners, and capital – [that] create self-organized order out of volatile, uncoordinated, confusing, conglomerations of countless different enterprises and individuals, narrowly pursuing countless picayune opportunities and their own interests. (Jacobs, 2000: 105–6)

But Jacobs diverges from Smith on the meaning of this process: instead of seeing the value of this self-correcting economy as a generator of creative diversification, the economics profession came to herald economic specialisation, which, for Jacobs, is to promote 'old work' efficiency (2000: 106–7).

For our purposes the most interesting part of Jacobs' legacy is the penultimate chapter of *The Nature of Economies* entitled 'The double nature of fitness for survival' (Jacobs, 2000: 119–32). Survival in ecosystems is not just about the process of immediate reproduction – competitively procuring food, shelter and other materialist and non-materialist needs – but also cooperatively preserving the wider habitat, the encompassing material world. Jacobs gives the examples of large mammals (lions, elephants, chimpanzees) with more than enough power to destroy their habitats but that have developed behaviours to prevent this. The basic method is to restrict time allocated to hunting and foraging (for example, by sleeping, 'play', grooming, etc). Jacobs asks the question 'whether our species has inborn traits that restrain habitat destruction' (2000: 125), and answers by speculating on the efficacy of several ancient cross-cultural restraints on habitat destruction such as awe of Nature (religions) (2000: 128–9) and 'corrective tinkering and contriving' (2000: 129–30). The latter, in particular, is integral to city economies, but therein lies a problem. This key urban mechanism is only about 10,000 years old, not enough evolutionary time to become an inborn corrective (2000: 131). Neither Smith's 'invisible hand' nor Jacobs' Commercial ethics syndrome offer any long-term constraining order.

So, we can accept that humans are part of Nature but that does not mean our species' particularities are not delinquent when it comes to habitat maintenance: the Plantation Age is our track record in this respect. And the worrying aspect is the nature of our species' particularity when it comes to the work of reproduction. As both Smith and Jacobs (1992: xi) observe, trading sets us apart from all other species. The latter use a local environment (a territory or trail) and this defines their habitat. Trading enables humans to exploit non-local environments for their reproduction, again epitomised by remaking of Nature through the Plantation Age. The key point is that the non-local sector of humans' habitat is beyond immediate experience. Hence it is difficult to see what ecological corrective mechanisms could develop whatever the time frame. For two centuries the human habitat has been global, although global-level effects may be much older than that. As we have seen, Jacobs postulates the idea of a new Age of Human Capital as the 'second creative age' (Zipp and Storring, 2016b: xxxiv), grounded in urban ingenuities whereby cities are available to generate future solutions to humanity's predicaments. And this is important, the standard 'Green' anti-growth arguments – curtailing city development – will make things worse by producing a world of stagnant cities, economies of old work using old technologies continually polluting more and more. Hope lies with vibrant cities: the crucial aspect of cities is that their dynamism lies in import replacement, which is ecological 'self-fuelling'. This is a process of continual localisation: in current Green parlance, it is decreasing humanity's footprint on its global habitat. Thus we complete this description of Jacobs' legacy where we started, on the nature of cities: 'The city is an invention for maximizing exchange and minimizing trade' (Register, 2010: 218). Therein lies the hope that by understanding cities through Jacobs' oeuvre, we can develop fresh meaningful ways of tackling the climate emergency.

Transcending Jacobs: interrogating her 'urban demand'

Hope is a precious commodity. We have presented a Jane Jacobs legacy that we think will help address anthropogenic climate change. The first question is, how is it to be done? This is very difficult to answer – it is the subject of remaining chapters – but how it is not to be done is clear enough. Near the end of her life Jacobs gave a lecture at City College, New York, where she proclaimed:

> I don't want disciples. My knowledge and talents are much too skimpy. The very last thing I would want is to limit other people with minds of their own ... we need unlimited

independent thinkers with unlimited skepticism and curiosity.... I am serious about not wanting disciples. (Jacobs, 2004b/2016: 458)

This is not an affected modesty of her later years but rather provides a strong steer of how she desired her legacy to unfold. She was an ever-curious independent thinker who encouraged curiosity in others to always challenge conventional thinking. In this spirit we identify a crucial lacuna in her economic development theory that directly impinges on our use of Jacobs' work to inform anthropogenic climate change.

This concerns our specific focus on the balance between supply and demand in the key development mechanism that Jacobs (1970: 150) famously claims to be 'a process of immense, even awesome economic force'. Yes, awesome: we will argue that these short-term growth spurts of cities create the urban demand that collectively has culminated in the existential climate change threat. Just like many other writers, however, Jacobs is strong on understanding supply but much weaker on demand.

Let us interrogate specifically how she describes the process. Starting with import replacement, this new local supply source satisfies an existing demand. The next step is import shifting, new additional imports of two sorts, goods to increase local production and goods for new local consumption. There is no explanation for why there should be any urban demand for these new imports. Jacobs' language is instructive here; her preferred verb is that cities 'earn' these new imports simply because they have replaced old imports. This is a sort of 'filling a vacuum' type of argument with demand as a pre-existing abstract economic space waiting to be filled. Thus in her discussion of the earliest cities replacing food from hunter-gathers by new agricultural production, she argues such cities were 'able to import other things in place of the wild food' (1970: 147). In other words, shifting demand remains as a given: 'In place of unneeded food imports, (cities) can import other things – a lot of other things'. A similar argument is used for modern cities: in the case of Los Angeles during its mid-20th-century urban spurt we are told the city 'could actually afford "extra" imports, not imported before' (1970: 154), wherein there was changing composition, 'certainly in quantity and probably in variety' (1970: 165). This form of thinking is confirmed when she presents her model diagrammatically where a city is said to be 'earning a sizeable quantity and diversity of imports' (1970: 255).

Could import replacement happen with no extra consumption of new goods so that urban demand stays constant – import reduction instead of import shifting? A city makes new things (supply – import replacement of an existing demand) but why an additional change (new demand enabling import shifting)? Jacobs provides many examples showing this does indeed happen, yet she does not explain why and how import replacement automatically induces import shifting. Put simply, Jacobs deals with conditions to facilitate additional production (customised imitation) but not additional consumption. This is because Jacobs' economics is built firmly on the process of work, and therefore her analyses are inevitably about production and supply. This extends to her philosophical studies where she interrogates 'working life', explicitly ignoring other aspects of living (Jacobs, 1992). Thus, in Jacobs' awesome urban spurts we are left with supply/production as process and demand/consumption as conjecture.

But we do not have to leave it there. What Jacobs is dealing with are urban agglomerations wherein there are economic clusters of related industries. The standard definition of a cluster given by Porter (1998: 197–8) is as 'geographic concentrations of interconnected companies, specialised suppliers, firms in related industries, and associated institutions (for example universities, standards agencies and trade associations) in fields that compete but also cooperate'. Note that this includes more than just the firms so that clusters are constituted as complex amalgams of supply and demand with new markets being generated alongside new productions. It is this basic agglomeration theory that we have to add to Jacobs' theory of economic development. In fact, Jacobs (1970: 235–47) alludes to cities creating demand in her discussion of differentiated production, and this becomes explicit in her incomplete historical narrative where 15th-century pioneers, printers and publishers are able to find new markets for their new products. Vibrant urban agglomerations encompass innovative processes, creating both new production and new consumption. Thus we can argue that there is always a latent urban demand – the attraction of cities as places of new consumptions – from the earliest cities through to modern mega-consumption, the mass urban demand promoted by the Advertising-Big Data-Social Media complex.

Linking Jacobs: some unlikely fellow travellers

We finish this chapter by resolving a problem we have ourselves created. By focusing on the oeuvre of a single person we have tended to isolate her work from other similar scholarly efforts. There is only one Jane Jacobs in the literal sense of her specific life experiences, reactions, interpretations and practices, but other scholars and activists have produced key ideas that relate to the various segments of her oeuvre. There are many radical thinkers whose work intersects with Jacobs' ideas in some quite unexpected ways. This intersecting of different situated knowledge can add further substance to Jacobs' thinking, providing perspectives perhaps beyond her ken. We use the cross-cutting of ideas to continue the momentum of serious unthinking of conventional social science thinking.

It is always thought-provoking when scholars from quite different starting points come to similar conclusions. In urban studies a classic case is that Jacobs (1970: 50) and Castells (1996: 386), coming from completely different theoretical positions, both assert the rare proposition that cities should be understood as process (unfolding) rather than places where things happen (in situ) (Taylor, 2006: 287). Such convergence acts as a sort of knowledge triangulation, and provides added confidence to contrarian ideas. The numerous intersections with Jacobs' ideas we present are offered in this spirit (see Froy, 2018, for a different set of intersections). In her local campaigns, in both New York and Toronto, Jacobs was famous for her kitchen table discussions with colleagues to map out ideas and strategies. For our purposes we need to take her out of this domestic/community milieu and place her where she can be at the centre of more global ideas and strategies. Figuratively this can be imagined as a watering hole for thinkers, Jacobs Bar – in the 18th century it would have been a coffee house – where scholars from different disciplines and political persuasion meet to discuss and compare ideas, and to plot protests. As a slayer of shibboleths, a thorn in the side of people who are certain, but ultimately offering humanistic positive thinking, Jacobs is the ideal bar owner to host contrarian discussion and actions.

We find that in Jacobs Bar, the seemingly odd, not to say outlandish, ideas and concepts that Jacobs developed are found to be not so idiosyncratic as they first appear.

- *The 'cities first' thesis* seems to have been universally rejected by archaeologists, and therefore this is an unlikely source for knowledge intersections. However, as well as supposed 'archaeological reality' (Smith et al, 2014) there are also 'archaeological illusions' (Pauketat, 2007). The latter more critical tradition does provide an intersection with Jacobs' idea of cities being directly implicated in rapid prehistoric change. Pauketat (2007) calls this the 'X factor' where a mixing of local and foreign practices create new traits. Originally derived to describe cultural change, he extends it to cover political change and brings cities into the argument (2007: 197) as 'the X factor of urbanization' (2007: 186). Furthermore, in his seminal work on Cahokia, Pauketat (2004) describes an economic basis for rapid urban change that has been interpreted as a 'perfect' replication of the Jacobs process (Taylor, 2013: 143).

- *Expect the unexpected* is the odd advice Jacobs offers to researchers if they are to engage in new knowledge building. This is an example of her distrust of conventional certainties and those 'experts' who hold them. But in studies of climate risk, uncertainty is endemic and therefore has to be integrated into decision-making (O'Brien and O'Keefe, 2014: 62). More generally, according to Funtowicz and Ravetz (1991) we are living in an era of exceptionally complex problems that require 'post-normal science' awash with uncertainties, and requiring rethinking of old 'certainties' like progress, modernisation and efficiency (Sardar, 2010).
- *Markets without capitalism* does not seem to make sense, and is therefore in urgent need of triangulation. And this comes from Fernand Braudel's (1981: 23–6) framework for his world history based on three levels of activity: everyday life at the base, commercial markets in the middle, and capitalism at the top, the latter including both states and corporations. Clearly the latter two levels relate to Jacobs' Commercial and Guardian work syndromes. Immanuel Wallerstein (1991: 206) interprets Braudel's framework as 'markets versus capitalism' so that, for him, Jacobs' markets without capitalism actually represent socialism. As an exemplar of Jacobs' political perplexity, praise for monopoly at the expense of competitive markets within capitalism is a key thesis of Peter Thiel's (2014) lesson for today's aspiring capitalists.
- *Poor cities need one another* is the opposite of trickle-down development economics. This is exactly the same message of the radical dependency school led by Gunder Frank (1969) who argues that the unequal relations between rich and poor countries keep the latter poor, a process he famously called the 'development of underdevelopment'. The practical implication of this has been argued by Samir Amin (1990) as the 'de-linking' of countries and cities from the world economy to enable a fair rebalancing of their local economies with like economies.
- *Debunking the 'Industrial Revolution'* has been a common topic since late 20th-century de-industrialisation of rich countries made equating 'modern society' with 'industrial society' problematic. But Jacobs going back a thousand years? The transition to modernity and/or capitalism is pushed back to around 1450 in Wallerstein's (1974) world-systems analysis and Janet Abu-Lughod (1989) goes back to the 13th century, but Jacobs takes it back to the beginnings of Europe's 'commercial revolution' described by Pirenne (1925/1969); this is empirically justified in Taylor (2013: 223–8, 240–3). Jacobs finds the first examples of differentiated production in the 15th century, which does coincide with Wallerstein's transition to a modern world-system.
- *The curious idea of a Plantation Age,* that Jacobs says we have been living through, is brilliantly portrayed by James Scott (1998), where

he compares the modern city planner's practices to the order created by scientific forestry with its logic of uniformity and regimentation (1980: 140–1, 184). Later he compares the city modernists as akin to a 'Taylorist factory' (1980: 348). The disorder of the growing city counteracted the state's need for legibility, which is consistent with Jacobs' Guardian syndrome, and leads to broadly anti-city practices to rule these unruly places. Of course, the greatest modern plantation builders were the communist states in the 20th century; from Lenin to Mao, they curtailed city growth and thereby stymied economic development in a promotion of old work, plantation-style (Davidovich, 1974: 627–9; Lin, 2002).

- *Present condition and future change* are treated by Jacobs as an urban-led transition, possibly requiring a societal bifurcation in escaping from plantation thinking. The latter includes 'industrial' mass production, and Manuel Castells (1996) has a similar formula of moving from an 'Industrial Society' based on spaces of places to a 'Knowledge Society' based on spaces of flows. And for Wallerstein (1991: 146–7) we are entering a period of *kairos* – decision time for humanity – when the modern world-system necessarily faces bifurcation, creating a different new society.
- *The reckoning of traditional economic thinking* has become a key element in contemporary scepticism of elites and their way of seeing the world. Contemporary globalisation has pushed cities to the fore as nodes in economic flows in contrast to traditional economic models of 'national economies' (Taylor et al, 2010). More generally the state-centrism of the social sciences has been widely critiqued (Agnew, 1993; Taylor, 1996b), but there has been no direct overlap with Jacobs; for instance, trade continues to be studied as 'international' despite growing proportions being intra-corporate. 'Free trade' remains a shibboleth despite some Green challenges and populist reactions.
- *Economics as ecology* is Jacobs' (2000) most profound engagement with economics and is used by Gibson-Graham (2008: 625) as a particularly creative example of knowledge making by bringing together different fields to generate ideas that could not otherwise be known. The resulting complex ecological thinking is identified as a contribution to finding alternative new economic possibilities outside capitalism. More generally, Jacobs' promotion of innovative small-scale economic process is wholly compatible with Gibson-Graham's envisioning of alternative economies to capitalism.
- *The reckoning of traditional party systems* is similarly widely appreciated, especially the demise of Centre-Left parties, the main legitimating element of liberal democracy. In the USA Packer (2013) calls this

stark new world of winners and losers 'The Unwinding'. The rise of populism in recent elections confirms the need to include Jacobs' political dynamics in electoral analysis (Taylor, 2017c). On the other hand, Jacobs' introduction of dynamic stability invites basic economic thinking, both radical, such as creative destruction (Schumpeter, 1975) and Massey's (1984) 'geological' layering of urban development, and conventional, such as product cycles (Vernon, 1966), into political analysis.

- *Reconnecting cities to Nature* has taken several forms, but most have focused on the energy efficiency of compact, dense urban settlement. Hence, 'New York is the greenest community in the United States' (Owen, 2009: 2). This statement was meant to shock because cities traditionally are viewed as the obverse of Nature, places where ecologies are obliterated by humans. Beyond simple density effects, we now find Jacobs' city process ideas becoming commonplace: learning in cities from environmental problems to produce innovation (Brugmann, 2009: 33), and the changing behaviour within cities reducing population growth (Brand, 2010: 51).
- *Holistic bottom-up knowledge*, which is what Jacobs Bar is all about, is critical to dealing with the inevitable 'produced unknowns' in adaptation to disasters (O'Brien and O'Keefe, 2014: 99). Post-event, resilience should not be about 'bouncing back' to a pre-disaster situation but rather 'bouncing forward' as a reaction to a changed reality, building a social transformation (2014: 138–9). This is recognised by the IPCC (2012). Piecing together a new world from a destroyed old world is dependent on local knowledge of survivors working in a holistic manner dealing with all aspects of change – economic, social, cultural – simultaneously, and requiring a very pragmatic politics: needs must.
- *Bottom-up knowledge and research*, finally, should not be ignored. As an activist, Jacobs is automatically orientated from below in challenges to authority. Although this did not have to be extended into her research methods, it was. In *Death and Life* she refers to 'foot people', those she observed in the street, as 'collaborators in the research'; she is very clear on this – local people did not 'influence' her; they collaborated (Jacobs, 1992/2016: 277). This is entirely in the spirit of participatory action research where there are no 'subjects' in the investigation; rather, people are 'producers of knowledge in the research process' (Askins and Pain, 2011: 806).

And so Jacobs' ideas are not so odd or different as often assumed; she was an original thinker to be sure, but her works should never be quarantined. The list obviously reflects our particular range of research interests; we invite readers to find other fellow travellers beyond our ken. But in complete contrast, note that we have not mentioned organised complex systems in the list. This is because Jacobs is widely recognised as the social science pioneer in this crucial area. That is to say, it is much more than a cross-cutting involving a few selected scholars. Her position has itself become conventional as complexity studies are part of the mainstream of today's social science (Wallerstein et al, 1996; Byrne, 1998; Batty, 2013) and in relation to climate change (O'Brien and O'Keefe, 2014).

We have kept the discussion of intersections at a simple descriptive level just to illustrate wider intellectual compasses into which Jacobs' ideas can be fitted. But what none of these other critical thinkers possess is Jacobs' acumen in locating cities at the centre of economic development and as such, crucial to understanding anthropogenic climate change. While keeping cities to the fore, the intersections have provided us with a broader grounding to our unthinking. We are more fully equipped for critically challenging key certainties held within mainstream social science. In the next chapter we take up Wallerstein's (1991) task of unthinking social science through its pretentious modernism, which is where Jacobs' radicalism began.

3

Inside Out: Fourteen Antitheses Authenticating Cities

Introduction: unthinking a thoroughly modern discourse

Without a sustained input from a critical social science, climate change science and policy-making have been condemned to being thoroughly modern. Whereas future scenarios – posited, projected or predicted – take humanity into uncharted waters (literally for many!), mainstream thinking about how we got into this predicament and how we might get out of it have been severely constrained by an embedded modern mindscape. Hence the need for a dose of Wallerstein's (1991) 'unthinking':

> I believe we need to "unthink" nineteenth-century social science, because many of its presumptions – which in my view are misleading and constrictive – still have far too strong a hold on our mentalities. These presumptions, once considered liberating of the spirit, serve today as the central intellectual barrier to useful analysis of the social world. (Wallerstein, 1991: 1)

Our attempt at loosening these constraints involves confronting conventional modern theses on the framing of human activities in time (chronology) and space (chorography) with plausible antitheses, thereby pointing towards a different understanding of the 'anthropo' in anthropogenic climate change. For chronology this means developing a 'trans-modern' approach that brings the 'pre-modern' into play in order to understand the possibilities of a 'post-modern'. The result is a break from the modern progress myth of humanity moving ever forward and upwards with its concomitant faith in technology to deliver a safe future. For chorography this means thinking outside the mosaic world created by nation-states that frame modern actions, economic and cultural, as well as political. Thus we break with modern state pre-eminence and its concomitant faith in good governments delivering a safe future. We view both dimensions of conventional thinking – progress/technology and state/policy – to be severely problematic: modern sinks dragging us into an abyss.

Our critical concerns are overtly manifest in the mainstream of climate science and policy-making, as represented by the IPCC and UN Climate Change Conferences (known as COP from 'Conference of the Parties' to the 1992 UNFCCC; the meeting in Paris in 2015 was COP 21) respectively. Quite overtly, these two remarkable global institutions epitomise state framing in both the science and policy-making. States select their national scientists for the IPCC whose work and publications, required to be 'policy-neutral', provide the main knowledge input to the COPs, where national policies are negotiated to combat global climate change. This modern chorographical basis of contemporary climate science and policy-making begets its twin modern chronological root of progress through technology. This is reflected in the different treatments of production and consumption in the work of the IPCC. In a very instrumentalist conception of the state that is employed in climate science and policy-making, the emphasis is on managing supply rather than curtailing demand (O'Keefe et al, 2010). COPs are largely about negotiating carbon emissions. The IPCC supports this focus through skewing the knowledge input in the direction of production over consumption.

This explicit production bias is clearly illustrated in Table 3.1, where the search results for selected words are listed from the IPCC's key 'state of the art' pronouncements, their five Assessment Reports since 1990. Here mentions of 'production' beat 'consumption' at a ratio of 2:1, but lower down the two lists the differences become overwhelming: in terms of economic sectors, references to 'industry' and 'manufacturing' far outstrip 'retail' and 'shopping'; in terms of policy approaches, 'technology' is out of sight compared to 'rationing' which is effectively off the radar (policy-neutral?). These reports do not say much about the nature of the society that is creating climate change, but there is a stark difference in use of the two abstract descriptions of that society in Table 3.1: it appears that while 'industrialisation' is an accepted part of the texts, 'consumerism' simply is not. The latter's frequency of just two within the many millions of words by several thousand IPCC authors is simply astounding to anyone vaguely versed in debates on the human contribution to climate change.

Table 3.1: Supply and demand: search results from the five IPCC Assessment Reports

Terms largely related to production		Terms largely related to consumption	
Term	Frequency	Term	Frequency
Production	49,500	Consumption	24,800
Industry	44,000	Consumers	3,840
Technology	39,100	Consuming	1,120
Producing	6,120	Customers	596
Producers	5,220	Retail	357
Manufacturing	4,950	Shopping	75
Manufacture	1,290	Rationing	9
Industrialisation	728	Consumerism	2

Source: Taylor et al (2016); produced from www.ipcc.ch/search/searchassessmentreports.shtml

Obviously supply and demand, production and consumption, are related pairings of single processes – you cannot have one without the other. The question is, for both theory and practice, how is each part of the pairing handled in terms of balance? Table 3.1 suggests a severe imbalance in mainstream scientific input into policy-making. This may well be a realist acceptance of modern politics; more consumption will likely create more support for ruling governments, so it is production that has to be made more climate-friendly, leaving consumption to carry on regardless. But whatever the reason(s), this situation is of particular relevance to our work because the narrative we develop in the next chapter strongly suggests that it is the generation of demand that is key for understanding anthropogenic climate change. Specifically, we identify urban demand as the vital mechanism of macro-social change so that cities, from their ancient origins to today's mega proportions, are directly implicated in creating exceptional levels of consumption, thereby inducing climate change. In our long-term story, the commercial world of cities is considered more important than political worlds of empires and states.

To explore this alternative chronology and chorography requires a critical assault on the modern thinking that believes that states, through intergovernmental negotiations, can solve an unfortunate and unexpected consequence of the 'Industrial Revolution'. We identify 14 conventional theses underpinning this modern mindscape against which we pit antitheses that provide arguments to help us think in a completely different way, towards a trans-modern discourse of cities encompassing pasts, presents and futures. The main substance of this chapter consists of presentations and explications of these 14 theses/antitheses.

Thesis/antithesis

This specific adversarial format has been chosen to demonstrate our critical approach because it emphasises an unremitting change of course from the timings and spacings of the modern mindscape. The variety of positions we take covers a lot of scholarly ground; most of the arguments are to be found in existing literature, and we provide these contrarian sources so that particular arguments can be followed up in more detail than we can provide here. Our contribution is to bring them together as a sustained critique. We have ordered them in a sequence that provides a coherent stream of thought, although in no sense does this indicate a ranking of relative importance. We begin with three arguments that provide the basis for a new way of presenting macro-social change; this leads to five arguments introducing alternative chronologies and chorographies; and these are brought together as four arguments engaging with climate change science and policy.

Each of the arguments is presented in the same manner to aid comparison and ease combination. After stating each conventional thesis and our antithesis there is an explication where we justify our position and link to its origin and to debates in the literature. Each thesis is treated as a 'given' in the sense of being conventional thinking within the modern mindscape, not uncontested, but widely accepted. This allows the focus of the discussion to be on the less accepted minority idea, the antithesis opposing the convention mindscape. The explication is followed by selected ramifications that point towards the next thesis/antithesis.

Sources of macro-social change

I: Cities over states?

Modern thesis. Transformative societal change is engineered through states because they are the prime units of collective human activity.

Antithesis. Innovative human activity is generated in and through cities and therefore they are the critical entities that create transformative social change.

Explication. The contrast here is very basic in terms of both time and space: what constitutes vital change, and how is the agency of that change spatially organised?

To answer the first question we use Braudel's (1972) concepts of time where he contrasts short-term history focusing on political events (*histoire événementielle*) with long-term history focusing on social structures (*longue durée*). Modern history has been dominated by the former, with discourses about states, their successes and demises, the rise and fall of empires, all pivoting on successions of key dates. This largely describes political change, which Jacobs (1992) understands as 'Guardian' agency where its zero-sum games create volatile geopolitical worlds of shifting winners and losers. From a long-term perspective, such interstate relations create international configurations that are historically ephemeral, important for the elites changing places, but not for the everyday lives of the majority of the population that continue much as before (that is, *longue durée*) under varying masters.

For the spatial organisation, we use Castells' (1996) concept of social space where he contrasts spaces of places with spaces of flows. Geopolitics is about territorial competition, altering spaces of places that are states. In contrast, spaces of flows are focused in and through cities where long structural change is enabled through commerce in its broadest sense, encompassing production, consumption and distribution (Jacobs, 1992). It is the long-term effects of Schumpeter's (1975) 'creative destructions' in cities, not the immediate military destructions by states, that define transformative material change (Jacobs, 1984). Hence cities are much more resilient than states, as reflected in the fact that most cities across the world are very much older than the states that currently encompass them.

Braudel (1972) emphasises that both the histories identified previously are important for understanding a full picture of change. Jacobs (1992) argues that both Guardian and Commerce agency is required for the reproduction of society. Castells (1999) belatedly realises that both spaces of places and spaces of flows are complementary components of social spatial organisation. To this we would add cities and states as both integral spatial units for our thinking. Therefore it is not a matter of which is right and which is wrong in these conceptual pairings; rather, it is a pragmatic choice dependent on the purpose of the chronology and chorography being created. The conventional thesis emphasising events, Guardians, spaces of places and states has served modern needs for good and ill over the last few centuries. It is our contention that its antithesis combining structure, commerce, spaces of flows and cities is required for transmodern thinking.

We contend that in terms of anthropogenic climate change, the modern chronology and chorography ultimately leads to a negative-sum game, only losers and further losers. Cities are the prime unit of human activity for countering this existential predicament.

Ramification. This antithetical argument has provided the conceptual toolkit we use for understanding chronology and chorography, and specifically identifies the critical importance of cities. Practically, it specifically directs us towards the need to focus on relations between cities and states in building alternative chronology and chorography for a trans-modern mindscape.

II: Cities created states?

Modern thesis. The state evolved out of chiefdoms; as the latter became increasingly complex, they generated additional political functions that culminated in state formation.

Antithesis. The dense peopling of cities generated conflict and the consequent demand for order was satisfied by inventing city-states; warfare among the latter created multi-city territorial states (empires) by conquest.

Explication. The thesis is an archetypal modern chronological argument that posits a simple evolutionary sequence. Building on the scientific reputation of Darwin, the idea of change as evolution has diffused beyond its knowledge field to be an easy means of designating causal relations. Thus an existing social institution is traced backwards to find less complex social forms until a simple origin is found. Initial Social Darwinism, with its racist overtones, was severely critiqued in the early 20th century, but the methodology survived, particularly in archaeology, wherein Gamble (2007) has provided a powerful critique.

In contrast, we argue that states, as demonstrated by their origins, are indelibly linked to cities. The conventional argument of the political evolution of states from class-less societies through increasingly complex class relations in 'chiefdoms' to finally create states has been contested by Smith (2003) and dismissed in detail by Yoffee (2005) and Pauketat (2007). Yoffee's counter-argument illustrates cities as generators of transformative social change: states are invented in cities. The coming together of people for reasons of commerce produces a dense and varied cultural demography in which social relations inevitably become fraught. State-making in cities is the solution to this urban conflict. Thus the city-state is the initial state form and is indexed by the building of city walls: there are typically a number of centuries between commercial city origins and conversion to state rule (Taylor, 2013). War-making between city-states produces winners and losers, thus creating traditional empires, states encompassing many cities, in a geopolitics continuing into the modern era.

What we are doing here is replacing a simple evolution theory by a demand theory of state origins: they were constructed to solve an urban need for order. The result is a new governance structure based on coercion; pacification of large territorial spaces involves the 'taming of cities', often indexed by the dismantling of walls of conquered cities. The key effect is relative loss of city autonomy so that the fruits of its commercial activities are taxed to pay for its own military subjugation. However, within territorial pacification, cities can still carve out their spaces of flows, often prospering by supplying the exorbitant demand emanating from imperial capital cities; grown large on tribute, they became the mega-cities of the pre-modern. This unequal city–state relation is typical before the modern period (Taylor, 2013).

Ramification. City–state relations are paradoxical. The awesome power of cities as world-changing institutions is illustrated by the invention of states but which then impinge on that power. We continue by exploring the inherent power of cities in macro-social change before returning to city–state relations under conditions of modernity in thesis/antithesis VII.

III: Cities are social development?

Modern thesis. Cities are outcomes of a wider social development that has produced places of dense activity we call urbanisation.

Antithesis. Cities are process, constellations of myriad urban networks that create social development.

Explication. In most modern scholarship cities are products, specific places, generated by more fundamental processes. Thus, typically, industrialisation is considered in some way to have 'caused' or at least led to modernity's historically unprecedented levels of urbanisation. This is particularly explicit in a Marxist approach with its class-based historiography encompassing the previous chiefdom/state thesis. This leads to a focus on cities as product, prioritising supply/production over demand/consumption. However, Harvey (2014, 2015) has recently moved towards our antithesis with an emphasis on circulation for the realisation of surplus value through property that suggests, at long last, geographical scholarship is searching for a theory of demand to underpin people's consumption.

But cities have not always been considered as simply outcomes; in our argument they are themselves processes, myriad networks creating dynamic mechanisms of change. Here we follow Jacobs (1970) and Castells (1996) who, as previously noted, independently both insist on cities as process. This position is much clearer in a trans-modern argument where cities are closely linked to civilisations.

Although in the modern perspective cities are viewed as subordinate to states (urban places within national territories), changing the context to civilisations elevates the role of cities. Because cities and civilisation are indelibly linked – a civilisation presumes the existence of cities – we are confronted with a chicken and egg conundrum. By thinking of cities as process they become prioritised as the 'egg' in incubation of civilisation. Civilisations are a consequence of cities as places of multiple innovations: their invention of writing creates history and their invention of states as empires provides the main subjects of history, an affirmation of the previous antitheses. In modern parlance, cities are development (Jacobs, 1970).

Ramification. Because cities-as-process is so demanding, its world-making potential requires a materialist rethinking of chronologies and chorographies; specifically, major societal changes commonly labelled as 'revolutionary' in modern discourse need fresh investigations (Gamble, 2007).

The key contribution to our unthinking by these three antitheses is to confirm cities at the centre of our discourse. With most people living in cities and using cities in their everyday lives, a common familiarity has left the incredible power of cities lost in plain sight. We have to rediscover this power in order to recognise how cities are essential to both understanding and tackling the climate emergency.

Alternative chronologies and chorographies

IV: Cities before agriculture?

Modern thesis. First there was an agricultural revolution, and when this evolved sufficiently to create material surplus to support city work, there was a consequent urban revolution.

Antithesis. Cities are very demanding, not least for food, and agriculture was developed to meet this demand.

Explication. These adversarial positions represent the most keenly contested and controversial part of our argument, as previously noted (Smith et al, 2014; Taylor, 2015). Archaeology as a discipline has pursued an evolutionary approach to settlement changes linking the process to an increasing supply of food. Explicitly codified by Childe (1950) into two revolutions, first, 'agricultural' and then, 'urban', the earliest cities are deemed to have been created about 5,000 years ago in Mesopotamia consequent on new higher levels of agricultural productivity. Note that this argument conflates Braudel's (1972) argument from thesis/antithesis I: the term 'revolution' was originally applied to political upheavals, explicitly short-term events (*histoire événementielle*), and was only later used for the *longue durée* processes that created cities and agriculture. Over the years this ordering of two revolutions, agricultural then urban, has become normative rather than empirical, and thereby uncritically accepted including into the sustainability literature (Steel, 2008). The problem for this thesis is that evidence keeps appearing for much earlier urbanisations; growing numbers of cities in the wrong place at the wrong time (Taylor, 2012a, 2013).

The antithesis is Jacobs' (1970) controversial 'cities first' argument, long denigrated and dismissed by archaeologists, but with growing support by urban scholars (Soja, 1990, 2010; Taylor, 2012a, 2013). Early indications of urbanisations are found millennia before conventionally expected across the world; the celebrated example, used by Jacobs, is Çatalhöyük in Anatolia from about 8,000 years ago. Deriving from a combination of trading networks plus new production practices, these initial cities not only preceded agriculture; they were also the reason for agriculture. As successful early cities grew, the hunter/gatherer means of supplying food became increasingly inadequate, and thus agriculture was invented to solve the problem. Hence the massive contrast between the orthodox supply theory of urban origins and this demand theory of agricultural beginnings. It is a key example of cities generating transformative social change.

Ramification. Bringing cities into questioning conventional views on a key early chronology opens up the possibility of extending this thinking to modern rapid societal change. But before we pursue this we can extend our critique to the agricultural revolution itself.

V: *Cities need orchards first?*

Modern thesis. As well as preceding cities, the 'agriculture revolution' focused on a particular form of farming: grain-based production first, in the 'Fertile Crescent' based on wheat and barley, and a little later, in China based on rice and millet. Mesoamerica's important contribution of corn (maize) came later.

Antithesis. This is a very narrow focus on a particular type of plant domestication, one that also just happens to sustain the modern world. It neglects other domestications beyond traditional hearths such as arboriculture as practised in pre-Columbus Amazonia. This points to a very different 'first' agriculture.

Explication. This antithesis is even more shocking than Jacobs' outrageous cities first thesis just presented. Crudely put, this thesis/antithesis pits fields against orchards as the original plant domestication. To view this contest from a city perspective we start with von Thünen's (1826/1966) famous agricultural land use model. He postulated a single city with circular bands of different production satisfying the urban demand. As a farmer and economist working in a pre-industrial world – with horse-drawn carts taking produce to market – he postulated intensive plant farming (fruit, vegetables) in the first zone, with the more extensive grain farming in the third zone away from the city. Such an intensive farming pattern is later found in the location of horticulture as market gardening. Is the first ring of agricultural production next to the city in space also the first production in time? Since early intensive plant agriculture includes perennials – bushes, vines and trees – it makes sense for this easier domestication to be the earliest to respond to urban demand for food.

This is what is suggested by the recent revelations of cities and agriculture in Amazonia (Mann, 2011; Clement et al, 2015; Pearce, 2015). The traditional modern view of the Amazon basin as consisting of pristine tropical forest is being replaced by the notion of an anthropogenic Amazon. This consists of the remnants of civilisations destroyed by the disease pandemic resulting from European contact, only now being discovered through airborne laser scanning technology aided by carbon dating and the genetic study of plants. In a region that is home to many scores of domesticated plants, about half of which are trees, much of the forest consists of abandoned orchards that had once provided for many cities. For instance, the peach palm has been cultivated for several thousand years. Requiring very little human attention and being very productive, it can be consumed in multiple ways, from cakes to beers. With trees, the distinction between 'cultivated' and 'wild' is less distinct; plant domestication is likely to be more about relocation, concentrating productive trees to be more accessible, a demand-led process. This intensive production is as much plant management as domestication, reflected in a major environmental impact of these civilisations, notably the widespread creation of local zones of 'Indian dark earth'. The earliest is dated six millennia ago. A remarkable soil rich in nutrients, 'the result of human waste management in and around settlements' (Clement et al, 2015: 3), it is a sure sign of cities, urban growth and demand steering economic productivity. And all this agricultural activity is described in von Thünen's terms with intensity reducing with distance from settlements (Clement et al, 2015: 3–4).

The conclusion is that 'Amazonia is a major world centre of plant domestication' (Clement et al, 2015: 2). Similar remote sensing of presumed pristine tropical forests in Africa and Asia also shows evidence of cities and therefore agriculture. This is an alternative narrative emerging on urban and food revolutions, but does it only apply to the tropics? Returning to von Thünen's model, berries, nuts, peas and beans are found alongside grains in ancient diets in more temperate climes, for instance in the best known 'first city' Çatalhöyük in Anatolia some nine millennia ago (Hodder, 2006: 82). However, in this case there is no discussion of intensive agriculture, and emphasis remains on the conventional research agenda of more extensive cereals and animal husbandry. There is clearly a need to reintroduce von Thünen's land use model into research on such important early agricultural development (Hodder and Orton, 1976: 229–36).

Ramification. This discussion is included to present a long accepted modern thesis that is clearly influenced by the geographical origins of the scholarship that has produced it. There is a casual Eurocentrism often accompanied by arguments for particular pre-eminences in human development: Diamond's (1997) celebrated world history – incidentally, fulsomely praised by Jacobs (2004: 11) – is a clear statement of the modern thesis written as Amazonian debates were only just beginning (Mann, 2011).

VI: Cities generate the transmission of agriculture?

Modern thesis. Agriculture as a 'revolution' has been a tremendous success story spreading inexorably across most of the world through either diffusion of ideas or migration of farmers.

Antithesis. Is the choice between diffusion and migration? No, because the success story is down to cities. As settlement nodes in trading networks grew in size and became cities, their demand for food instigated the spread of agriculture.

Explication. According to Hassett (2017: 80), the most intriguing thing about the 'agricultural revolution' is that it was its 'runaway success', despite the evidence, from analysis of bones that the changes were hugely detrimental to human wellbeing. So why did this new way of life turn out to be so popular? And it surely was. From multiple-origin locations, agriculture spread far and wide. To be sure there were some reversions to hunter-gathering, possibly related to local climate changes, but these were specific exceptions from the general process, a persistent transmission of farming.

Archaeologists have long debated two means for the spread of agriculture: migration and diffusion. In the first process farmers moved out of origin locations taking their new way of life with them as they replaced earlier settlers. The second process has the movement of knowledge out from the origin location as surrounding populations copied the new agricultural methods. The debate has been subjected to fresh investigation with the development of DNA analyses of human remains: it can be shown whether a new agricultural population in an area is similar to, or different from, the prior hunter-gather population. If the former, the process is diffusion; if the latter, migration. In the classic case of transmission of agriculture from the Middle East to Europe the results are very mixed (Cunliffe, 2008: 89); Hassett (2017: 81–3) describes a complexity of outcomes with evidence for both processes in a complicated geography.

So instead of choosing between the two processes we need to understand this complexity. According to Marshall Sahlins' (2004: 1) analysis of 'stone age economics', hunter-gather communities constituted 'the original affluent society'. This was because their needs were finite and few and therefore satisfied through relatively little work, enabling much leisure time. Changing this social living to the much harder (more work, for less nutrition) world of agriculture is an immense turnaround. Such social transformation occurring relentlessly requires a very specific context found only through the complexity that is cities (Taylor, 2013: 112). Hence the spread of agriculture links to the invention of agriculture with the movement of commodities (trade), people (migration) and ideas (diffusion) all in the mix of evolving of trading camps and networks into rudimentary city networks (Taylor, 2013: 113).

Ramification. This antithesis has the greatest ramification for the overall argument of this book because it is here we find not only the creation of urban demand but also the first expression of its immense power. There is no political instrument coercing people to be farmers; they are responding to the first world-changing urban demand – for food.

VII: Cities transcend industry?

Modern thesis. Starting in one state, late 18th-century Britain, and diffusing across parts of Europe and North America in the 19th century, industrialisation created new national societies that are the first modern and/or capitalist societies.

Antithesis. As an economic formation, capitalism is transnational in space and 'trans-industrial' (that is, beyond one economic sector – manufacturing) in time; a transition to a capitalist world-economy occurred in cities through the 'long 16th century' (ca 1450–1650) to create the modern world-system in which we still live today.

Explication. This is about the meaning of the 'Industrial Revolution', the conceptual precursor and historical successor to the agricultural and urban revolutions in thesis/antithesis IV. There are two entwined debates involved. First, there is the challenge to state-centric thinking most explicitly expressed in world-systems analysis since the 1970s (Wallerstein, 1974, 1979, 2004). As such it attacks both Marxist and liberal acceptance of economic process coinciding with sovereign political territories, and advocates a systemic approach to capitalism that transcends political boundaries. In its embryonic form this transnational formation was predicated on cities (Braudel, 1982, 1984), and this has more recently been strongly supported empirically (Taylor et al, 2010; Taylor, 2013). The chronological effect of this revisionist thinking has been to trace capitalism and modernity back to before the so-called Industrial Revolution, specifically to the beginning of European expansion some two centuries earlier. In geography, this move has been thoroughly endorsed by Jason Moore (2014) on environmental grounds; it involves movement of flora and fauna (including diseases) between continents in what Crosby (2004) has termed 'ecological imperialism'.

Second, there is the recent conceptual delinking of 'industrial' from both 'modern' and 'capitalist'. For 19th-century scholars experiencing a new industrial world based on rapid technology advances defining human 'progress', it all came together as industrial being synonymous with both modern and capitalism. For instance, industrial society *was* modern society (within a modern state), to be contrasted with non-industrial and therefore 'unmodern' societies. For the latter to become modern, they would have to become industrial, which is how 'development' policies, both capitalist and communist, were framed in the second half of the 20th century. But when the original modern countries began to de-industrialise and remained rich while industrialisation became a feature of poorer countries, the conventional link between modern and industrial was unequivocally severed. Combine this with the world-systems critique and the concept of modern (and capitalism) becomes perforce trans-industrialisation.

Finally, the transition to modernity (and capitalism) alters the paradoxical relation between cities and states. With multiple states rather than a single over-arching world empire, economic elites were able to engage in a more equal relation with political elites, and this enabled cities to prosper as economic development moved from northern Italy to north-west Europe in the 'long 16th century' (ca 1450–1650). But increasingly these multiple modern states accrued many more functions than traditional imperial states, starting with borrowing mercantilist policies as pioneered by cities. We know the end result is modern urbanisation on a scale totally different from anything that went before.

Ramification. We have undermined the idea of industrialisation being the foundational turning point in making the modern world. This is in keeping with key previous arguments: our prioritising demand over supply, and the *longue durée* process over 'revolution' representing short-term (*histoire événementielle*) change. However, this position appears out of sync with climate change discourses that emphasise the importance of the 'Industrial Revolution'. This is addressed in thesis/antithesis IX in the climate change section.

VIII: Subsistence agriculture without cities?

Modern thesis. Worldwide, subsistence agriculture has been the foremost form of food production, both historically and in many poorer countries, up to the present.

Antithesis. Farming for exchange through urban hinterlands or networks has always been the primary form of agriculture; farming for subsistence represents a regression consequent on urban decay.

Explication. Local subsistence agriculture seems a natural starting point for evolutionary interpretations from rudimentary methods (slash and burn) for self-consumption to increasingly intensive and productive methods for wider consumption. The cities first theory completely overturns this since agriculture is invented in order for its products to be exchanged. Therefore subsistence agriculture is not positioned as 'not yet commercial' but rather, as formerly commercial.

As a product of urban demand, agriculture prospers or declines with its market in cities. In the limiting case of the demise of cities, agricultural villages will lose their raison d'être; they will be incomplete fragments of a past economic world. Jacobs' (1970) calls them 'orphaned settlements'. Agricultural skills are not immediately lost, but the work has to be re-orientated, to fall back on the only surviving consumption, that of agricultural workers and their families. It is not just the unavoidable reduction in quantity of production; there is also a crucial loss of urban opportunities for development creating an inevitable stagnation. For Jacobs (1970) they become 'by-passed places'.

Strangely, this part of Jacobs' cities first theory has not been subject to discussion in the literature. However, there is one intriguing corollary of this interpretation: regions of widespread subsistence agriculture become obvious sites for searching for, or expecting where others will find, early 'lost cities' (for example, Mann's [2011] 'humanized landscapes' in pre-1492 Americas; see also Clement et al, 2015).

Ramification. This is a beginning for developing a new chorography based on urban-based demand. Problematising the separateness of agricultural places and landscapes leads to questioning of the very idea of 'rural'.

IX: Cities encompass rural places?

Modern thesis. There is a critical division between urban and rural as a theoretical and practical distinction deriving from contrasting land uses that have created very different sorts of places.

Antithesis. The city process incorporates both urban and rural places in a singular dynamic.

Explication. The idea that urban and rural are fundamentally different social realms long precedes modernity, but with the latter's massive urbanisation, the rural has taken on a distinctive conservative role. In modern politics 'nations' are defined by their rural places (for example, English countryside, American frontier) irrespective of the degree of urbanisation. But this does not mean that radical scholars have not used this chorography, Raymond Williams' (1973) *The Country and the City* being only the most explicit example. Generally such work reinforces Childe's chronology and in geography prioritises place-content (settlement type, land use) over integrative process. Globally this resulted in an urban studies focused on the 'Global North' (urban systems research), with rural land use studies dominating 'Global South' research (development planning).

This spatial separatism has been challenged in different ways: for instance, empirically by Cronon (1991) and more theoretically by Amin and Thrift (2002). Most recently Brenner (2014), through his planetary urbanisation initiative, has argued that the urban is everywhere as a global functioning complex. This is an application of our previous argument considering city as process rather than place. It follows that the widespread reporting of the global population passing the 50 per cent urban threshold misses the point: the vast majority of the world's population have long been organised to meet the demands of cities. More generally, all urbanisation is reliant on populations beyond cities to supply in-migrants. This urban–rural functional link has been crucial historically because spatially concentrating people creates unhealthy places where death rates exceed birth rates; the 19th-century public health policies cut this link, but the immensely increased global urbanisation since then has continued to be supplied largely through rural–urban migration, most notably in China since 1980.

In addition, the process of city demand for food further undermines Childe's political language of revolutions ('industrial' as well as 'agricultural' and 'urban'). The enormous increase in large city populations by the end of the 19th century – Weber's (1899) new world of great cities – should not be deemed simply 'industrial' social change; rather, our cities approach reinforces Brooke's (2014: 480) argument that the key environmental trigger is immensely heightened urban demand becoming worldwide. For instance, the great urban expansion includes explosive city growth in the frontiers settled by English-speaking people as described by Belich (2009) (see also Taylor et al, 2010). The movements of people between and within continents generating growing trade in commodities such as sugar, tobacco, coffee and tea may appear superficially as workers moving between 'rural' places, but they were actually caught up in a single urban dynamic.

Ramification. Given a singular urban dynamic of social change, cities should be central to climate change science. For climate change policy this implies a change of emphasis from state supply-based solutions to interventions in urban demand.

The key contribution to our unthinking by these five antitheses is to place cities as instrumental in the key macro-social changes that are implicated in anthropogenic climate change. This is because how we think about change, especially origins, is a prerequisite for the necessary understanding to resolve the situation (Moore, 2015: 4; see also Angus, 2015).

Anthropogenic climate change science

X: Cities orchestrating anthropogenic climate change?

Modern thesis. Anthropogenic climate change started with the Industrial Revolution about 200 years ago, resulting from the continuously increasing use of carbon fuels in production.

Antithesis. Anthropogenic climate change has been happening for 8,000 years and was initially the result of land cover removal for agriculture to feed early cities.

Explication. There is little doubt about how mainstream treatment of anthropogenic climate change considers the 'Industrial Revolution': it is by far the key historical concept in this conventional thinking. We would surmise that it is the main reason for the production bias discussed earlier (see Table 3.1). It is fundamental to the modern chronology because it specifies the beginning of anthropogenic climate change.

Ruddiman's (2003, 2010, 2013) research on the constituents of climate change – and the changing levels of greenhouse gases – has resulted in a serious challenge to the Industrial Revolution starting point. His method is to chart changes in greenhouse gases as multiple cycles over many ice ages – 'nature in control' – and then search out anomalies in the period since the last ice age, which he attributes to 'humans in control'. His findings show an anomalous rise in carbon 8,000 years ago and an anomalous rise in methane 5,000 years ago. Combining these, he produces a new chronology for anthropogenic climate change involving two processes: a slow increase in greenhouse gases starting 8,000 years ago and a rapid increase in greenhouse gases over the last 200 years. Although the early rise is slow, he argues that it should not be under-estimated because of its longevity relative to the recent rise: he uses the tortoise and the hare analogy. The human side of Ruddiman's chronology is conventional: he accepts that the recent rapid increase is a result of the 'Industrial Revolution'; the early slow increase is explained as caused by the 'agricultural revolution', specifically, removal of land cover.

The most basic critique of Ruddiman's thesis is an empirical one: early populations were not large enough to have made the environment impact he posits. He counters this by citing increased estimates of both population totals (Gignoux et al, 2011) and the need for extensively large clearances in early agriculture (Ruddiman and Ellis, 2009; Kaplan et al, 2010). However, this debate is by no means settled, and our introduction of cities into the argument can substantially augment Ruddiman's position. By focusing on urban demand for food rather than rudimentary subsistence agriculture, we provide a completely different chorography, a geographical imagination of city hinterlands and networks as an alternative to simple demographic counts of subsistence farmers. Caused by the urban demand for increased production, land clearances are thereby predicated on a much more complex demography and economy. In this argument discoveries of ancient land cover removal for agriculture represent an initial urban ecological footprint. Thus linking Ruddiman's position to Jacobs' (1970) cities first argument generates a credible case for initial anthropogenic climate change being a consequence of early city process (Taylor et al, 2015; Taylor, 2017a). The initial land cover clearances required for provisioning cities by dryland cereal production is implicated in the rise of carbon emissions; subsequent wetland cereals production to provision ever-growing cities is implicated in the later methane emissions.

There are two important chronological questions that arise through bringing Ruddiman into our argument. First, an intriguing implication of Ruddiman's (2010: 95–105) research is that although we think of anthropogenic climate change as a bad thing, this is only so for the current phase of rapid rise. The long slow human effect of global warming before 1800 had been immensely positive for humans. It prevented the return of another ice age and thereby provided a unique climatic window of opportunity: a long period of stable environmental conditions that suited human material development. Second, his slow/fast anthropogenic climate change division does coincide with conventional identification with the 'Industrial Revolution' rather than our 'long 16th-century' transition to modernity/capitalism. Why the delay? Well, in the beginning the modern transition was in no sense global; its new urban demands were limited with respect to contemporaneous larger traditional empires, notably China. Additionally, the expansion into the Americas created a massive pandemic from 1500 to 1800 that Ruddiman (2010: 132–3) recognises as actually lessening human impact on climate; in our argument this is crucial because the decimation of urban hinterlands and networks across the Americas (Mann, 2011) countered the effects of urban growth elsewhere, thereby delaying a potential climate effect.

Ramification. There are two initial consequences of this merging of contrarian arguments: first, there is an urgent research need for modelling the early urban landscape to estimate increased greenhouse gas emissions; and second, through refuting the modern thesis we provide a direct challenge to the foundation of climate policy-making as currently conducted. Here we focus on the latter.

XI: Cities are so demanding?

Modern thesis. Anthropogenic climate change as a relatively recent phenomenon can be tackled by states through negotiations on reducing carbon emissions.

Antithesis. The starting point for tackling anthropogenic climate change is to understand that both early and late transformative alterations in climate have been generated by demand through cities.

Explication. What the 8,000-year chronology provides is an understanding of the power of cities, initially a long, slow growth of global urban demand followed by a rapid acceleration of that demand. The latter is continuing and quickening, which means that the situation is becoming more and more urgent. But this does not lessen the need for a full historical interpretation of the problem. Thus to concentrate our scholarship exclusively on 'carboniferous capitalism', as the modern burst of economic growth has been commonly called, is misguided. It is not the only way humans have caused climate change. And in any case, it is not the 'carboniferous' that is the root concern; it is the 'capitalism' as ceaseless accumulation, the modern economic system that can only continue to exist through continuously growing consumption.

The conventional modern chronology of only 200 years of anthropogenic climate change fits neatly with thesis I, the preeminence of the state. The result has been deployment of an elementary instrumental theory of the state that generates a simplification of a complex subject so that governments are able to do something practical (Scott, 1998). The focus of governments on carbon emissions makes short-term sense – 'keeping the lights on' – buttressed by neoliberal economic ideology with its market short-termism. There are minor concessions to thinking longer term in terms of government subsidies for non-carbon energy sources and carbon trading to effect territorial carbon budgets, but the politics and international relations remain trapped in a modern mindscape that cannot handle the complexity of the global predicament. This misconstruction premised on supply is clearly illustrated by the UK government's aptly named policy instrument, the 'Department of Energy and Climate Change'.

Back to basics: energy is produced as supply and consumed through demand. At best, policy that privileges supply is dealing with only half the energy system, but in our argument it is much worse than this: it misses out the crucial part of the system which is demand for energy predicated on the wider material demand generated through cities. Further, it lacks the links to the demand that cities are placing in terms of the 'rural' responding to climate change (O'Brien et al, 2011). This understanding comes from the 8,000-year chronology integrated with the chorography of city-centred flows. Territorial state energy budgets can only make sense in strict autarky where supply and demand are contained together. Any porosity through political boundaries constitutes outsourcing of energy and other material flow. Thus measuring territorial environmental footprints is fraught with misunderstanding: a product consumed in state A constituted by inputs from states B, C and D violates the spatial integrity of any bounded measure in a myriad complexity of flows (see Wiedmann et al, 2015, on consumption-based material footprints). In complete contrast, our cities' mindscape is constituted by flows, paths, routes, connections, chains, links and circuits, with boundaries having only a cursory presence as obstacles. Instead of territorial footprints there is an urgent need to research city 'net-prints', the demand power and scope of cities, as a basis for policy-making (Taylor and Derudder, 2016). This is broadly cognisant with current descriptions of planetary urbanisation (Brenner, 2014) that are going beyond territory literally by including the oceans in urban process. In terms of our urban dynamic argument, this is represented by city net-prints covering ocean fishing for food; for instance, fish from the North Atlantic to feed workers in the cities of northern Europe in the 'Industrial Revolution' era, and earlier supplying for religious needs in medieval cities across all of Europe. Today we live in an urban oceanic world of industrial fishing and ubiquitous plastic waste.

Ramification. Prioritising urban demand will necessarily problematise place-based policy initiatives and join with current environmental concerns for material flows. Going further, by bringing cities to centre stage we are forcing complexity on to the agenda; engaging with cities should always respect these settlements as the most complex of all human artefacts. But bringing cities into the argument has not always been accompanied by complexity.

XII: *Cities for green efficiency?*

Modern thesis. As specifically dense places, cities are the most sustainable of settlements, and their remodelling as smart green cities is the place-based urban way of tackling anthropogenic climate change.

Antithesis. Anthropogenic climate change has to be addressed as the result of cities as process; it is a matter of consumption, collective material demands through myriad urban networks.

Explication. These positions identify two key themes in our argument: in terms of chorography, the difference between city as place and city as process, and in terms of chronology, alternative views of technology in society.

Recent years have seen a surge of interest in green, eco, sustainable, compact, etc, cities. 'Green Cities', for example Masdar in Abu Dhabi, have been designed and built from new. Others, such as Aalborg in Denmark, have had the existing built environment refurbished along with new 'sustainable' developments (Joss, 2015). The underlying assumption is that we can build or refurbish cities as our way of 'solving' the climate problem. But this fails to recognise that cities are vitally processes, material networks at multiple scales leading to concentrated consumption, today at mega levels. Simply focusing on design of place diverts attention from our current lifestyle embedded in an economy of continuous growth. Such arguments are based on a belief that technology will find the solutions needed to address climate change. Certainly there is little to suggest that technology, in particular technology transfer, has addressed energy poverty in poorer countries (O'Brien et al, 2007).

The basic problem with relying on technology is that it cannot be separated from the society in which it is created. Modern technology is first and foremost modern. Treated as a specific practice, as humanity's saviour, it prevents transcending a modernity that is inherently consumerist. As part of an economic system that requires ever more consumption, technology's prime use is to generate ever more products. This logic requires built-in obsolescence accompanying ever-changing fashions, both generating new needs to fuel demand created by sophisticated marketing campaigns, enabling corporations to keep producing more and more stuff. But the reality is that we need an approach that requires us to have less and less stuff, which means a completely different economic logic. A sustainable approach to cities within a mega-consumerist economy is simply impossible. Planning one city at a time creates a landscape dotted with 'sustainable cities', which is a wholly inadequate response. Put bluntly, sustainability cannot be local. We need to think in a more sophisticated way, a geographical imagination, a new chorography, based on a holistic urban approach to changing behaviour and greening the economy (O'Brien and O'Keefe, 2014).

It should be noted that this position is not an anti-technology argument: better use of energy will be a necessary part of any holistic approach to tackling climate change, but it can never be sufficient. However, operating as the latest manifestation of the modern progress myth, technology is positively dangerous (viz geotechnology!).

Ramification. It is not just a matter of focusing on cities; it is how cities are understood that matters. Our continuing message is that cities are much more than a type of place; they are a process and moreover one with world-changing powers. In our current climate predicament where change is urgent, this can be a good thing.

XIII: Cities as new polycentric structures?

Modern thesis. There are new polycentric city-regional formations that are the necessary framework for global policy-making to engage with both economic competition and making a sustainable world.

Antithesis. Polycentric city-regions are historically ubiquitous and, having been central to the operation of city process as economic cooperation, are crucial for a necessary global transition.

Explication. Building on Gottmann's (1961) megalopolis concept linking US Eastern Seaboard cities from Boston to Washington DC, there has been a plethora of findings of such mega-polycentric urban regions across the world (Choe, 1998; Faludi, 2002; Hall and Pain, 2006; Harrison and Hoyler, 2015) that now feature in planetary urbanisation (Brenner, 2014). This globalisation of Gottmann's concept can sometimes appear as a celebration of size – better to compete, better to sustain – with implications of evolutionary inevitability. Its chorography can also be quite problematic, with an acute concern for spatial delimitations that betrays a territorial emphasis, sometimes with Scott's (1998) state-like easy simplifications. Despite the emphasis on planning in this literature, it can appear that we are advancing towards an urban dystopia, a new mosaic world of urban behemoths (Petrella, 1995).

Of course, cities have commonly clustered in successful regions of economic development, both cooperating and competing in dynamic innovative cultures. Historically, modernity has been built on three such multi-nodal urban regions: Holland in the 17th century, northern Britain in the 18th/19th century and the USA 'manufacturing belt' in the 19th and 20th centuries (Taylor et al, 2010; Taylor, 2013). Pre-modern, stretching at least from the Mesopotamian urban blossoming 5,000 years ago (Algaze, 2005a, b) to the late-medieval northern Italian urban region (Arrighi, 1994), this is how city process has operated most successfully (Taylor, 2013). And this process, modern and pre-modern, has been very much a bottom-up mechanism through communities and businesses benefiting from agglomeration and connectivity advantages to alter their worlds, usually in small ways, sometimes amassing into social transformative change. Intimations of this dynamism can be gleaned in the contemporary city process, for example in Lang's (2003) rich American 'edgeless cities', but the innovatory behaviours for survival in adversity, as in Neuwirth's (2006) poor 'Global South' 'shadow cities', are probably more relevant for the transformative change that is now required. Whatever, the key point is that in our uniquely 'urban century', city process is both more potent and more needed than ever as the locus for tackling anthropogenic climate change. Satisfying urban demand locally through an alternative form of development (that is, building on Jacobs' [1970] import replacement mechanism as localisation) provides initial hints towards realising a utopian vision of green networks of cities (Taylor, 2012b; Taylor and Derudder, 2016).

One final point – this evocation of bottom-up process is not simply the inverse of dependence on top-down state-negotiated policies; it is also the reverse of urban top-down politics: if mayors ruled the world (that is, back to city-states), we would definitely be travelling to another simple urban dystopia.

Ramification. This brings us back to the paradox of city–state relations. Our *longue durée* arguments are confronted by a current urgency. Certainly the bottom-up process we have described requires a bottom-up politics through which change is debated and navigated. Transition politics will necessarily be very different from modern politics, but it will still need operational political instruments.

XIV: Cities as contrary to Nature?

Modern thesis. Cities represent the very opposite of Nature; they are artificially built environments that overlay and destroy natural environments. Concrete versus fields: they are the most 'un-green' places on Earth.

Antithesis. Nature is an assemblage of ecologies of which cities are vibrant examples. Cities are built environments just like all other ecosystems; however, the builders of cities, humans, have evolved to become the most powerful species within Nature.

Explication. The modern thesis appears to simply state common sense: cities don't look like Nature is supposed to look. And this view is totally compatible with how scientists have approached contemporary climate change. IPCC Reports, in many ways, are a model of science. Peer-reviewed conclusions, from a shared paradigm of theories and methods, and most importantly a growing number of practising scientists, provide for evidence-based policy formulations to address the problem of climate change. What is the problem? The problem is the concept of Nature.

The separation of cities from Nature is a consequence of the development of modern thinking that places 'Nature' and 'society' in separate domains. Moore (2015) argues that at the heart of environmental concerns there is a Cartesian narrative (that is, derived from the ideas of the 17th-century philosopher Descartes) in which modernity, especially in its industrial phase,

> ... emerged *out* of Nature. It drew wealth *from* Nature. It disrupted, degraded, or defiled *Nature*. And now, or very soon, Nature will exact its revenge. (Moore, 2015: 5; original emphasis)

The climate emergency signifies that revenge.

Moore (2016: 1–6) offers a post-Cartesian world view in which the Nature/society binary is replaced by a world ecology that he calls the web of life. This is Nature as a whole, a realm of flows, dynamic relations of which human activity is but a part. For Moore, treating Nature as external is toxic: it is a means for making ecosystems to be solely organised for human needs: multifarious social activities – government, science, empire, industry, and so on – are each projects for organising Nature, including humans, for social ends. Rather, for instance, families, corporations, markets and cities are all natural forces within the web of life. With numerous other human institutions they represent the specificity of the human species, its sociality, resulting collective capacity and power. In the case of cities this amounts to what we term 'demanding the Earth'; it is adherence to the Nature/society binary that makes this possible.

All this gels perfectly with Jacobs' (2000) interpretation of city processes, not as analogous to natural processes, but as being actually natural. Thus dynamic cities and tropical rain forests are both examples of intricate flows of energy creating complex eco-systems.

Ramification. Jacobs (2000: ix–x) recognises that the notion of a unity of Nature including humans is difficult to accept for people viewing it from positions both sides of the Nature/society binary. Ecologists exhibit anger at what humans as 'interlopers' are doing to Nature, while all manner of modern practitioners (economists, politicians, etc) interpret human achievements as going beyond – improving on – what is found in Nature. Thus 'green' urban initiatives invariably remain binary. One exception is 'ecological planning' as famously championed by McHarg (1969). His exhortation to 'design with nature' encompassed an anti-industrial modernity ethos; a severe critic of 'power over nature', he proposed using natural processes in urban design practice that aspired to be 'steward of the biosphere' (Wenz, 1995). Wenz notes that McHarg did not always provide a 'complete ecology' in practice: humans in his holism are signified by the landscape, represented morphologically rather than in the complexity of city flows. But the Nature/society binary is recognised and largely overcome in his work. In contrast, the consequences of not dismissing binary thinking are profound; where Nature remains a separate domain, there is an external nature ripe for dangerous human projects.

The key point is that human reaction to the climate emergency requires more than 'green' urban reforms making cities more Nature-friendly; it requires an ecological reinvention of the city in Nature.

The key contribution to our unthinking by these five antitheses is to insert the nature of cities into the realm of climate change policy thinking. Past, present, and by implication, future climate change has been, is and will be decided in our cities. The future nature of our cities will determine the success or otherwise of anthropogenic climate change policy. In total, we have come full circle, starting by dis-embedding cities from state (I) to re-embedding them in Nature (XIV).

Political reflections

It is usual for a thesis/antithesis mode of argument to conclude with a synthesis. Our synthesis is presented in the next chapter as a trans-modern narrative on cities being so very demanding for over 8,000 years. Self-evidently this is a 'big picture' approach (with associated 'grand narrative') that is not always seen as legitimate in critical thinking; our defence is simply that anthropogenic climate change is a big picture topic. However, as Braudel's (1982, 1984) work has bequeathed to us, long-term history plus large-scale geography does not have to neglect agents of social change, hence our bottom-up ending to thesis/antithesis XIII. But we cannot finish our argument here. The 14 critical takes on conventional modern thinking are not intended as a set of academic exercises for rearranging social science research on cities and states; rather, they are intended to contribute to a fundamental mindscape break required for the immense, urgent human task of tackling anthropogenic climate change. Following on from thesis/antithesis XIII, our initial reflections focus briefly on political implications, on the hugely difficult task of bridging what Castree (2015) calls the 'knowledge–action gap'.

A key point to conclude with is that we need a trans-modern sensitivity in our political work. The chronology or chorography promoted previously provides a positionality that will inevitably add some modesty to our politics. In anthropogenic climate change there are uncertainties in terms of future physical changes, but these are dwarfed by the myriad possibilities for societal change resulting from the demise of modernity. The consequences are profound and unbelievably difficult. Wallerstein (1992: 561) has argued that in attempting to comprehend a 'post-modern or capitalist' future we are in a similar position to a 14th-century peasant trying to forecast our modern world. But we are where we are, and this is where the political action is: Braudel's (1972) *événementielle* from thesis/antithesis I. Multifarious short-term politics has been given their direction through the progress or evolution discourses of modern radical movements: practically the 'forward march of labour' and/or other oppressed categories, and theoretically the Marxist transition to a communist mode of production, to which we can now add Klein's (2014) 'unfinished business of liberation' for progressive climate change activism. But we have argued that these evolutionary, revolutionary and progressive arguments do not satisfy our trans-modern way of thinking. Further, there is a long tradition of modern bottom-up political movements degenerating into an alternative top-down elite politics. This is the opposite of the bottom-up processes envisaged in thesis/antithesis XIII. We never said this would be easy.

4

Reset: Anthropogenic Climate Change Is Urban, not Modern

Introduction: trans-modern scholarship

The last chapter made the case for unthinking in a trans-modern manner; here we implement this approach. But developing trans-modern scholarship is fraught with knowledge difficulties. In effect, our take on anthropogenic climate change is a provocation to modernity's innate superiority complex. We are downgrading our modern world from ultimate historical climax to just another historical interlude. This changing macro-positionality means we are basically searching for a sort of 'self-reflexivity' at a societal level, moving towards a global ecological holism that Jacobs understood many years ago:

> The fragile ecology of a city and the fragile ecology of the Arctic stand or fall together. Because a mounting problem has gone unsolved in cities, in deference to the status quo, the outermost wilderness is finally threatened. (From a speech by Jane Jacobs in 1970, quoted in Zipp and Storring, 2016a: 203)

We tackle this transcending of modern sensibilities, as indicated by our title, by transferring the 'anthropo' in contemporary climate change from industrialisation, as specifically modern, to urbanisation, as generic focus: modern people are not the only humans to influence planetary climate change. The chapter consists of two arguments, the first developing and presenting a narrative for anthropogenic climate change, and the second making trans-modern interventions through consequent new chorological and chronological framings.

As described in the last chapter, at the heart of the modern mindset there is grand narrative combining a progress (development) chronology with a mosaic (states) chorography. This can be challenged in two ways, by eschewing grand narrative as legitimate reasoning or by constructing a contrary narrative. In practical terms we think the former is the risky option for radical social change at the macro level because it leaves a pedagogic vacuum. Thus we confront the modern narrative with an alternative narrative customised for humans' role in planetary climate change. This combines a generic process – urban demand – with historical specifics in a story that importantly includes, but does not extoll, modernity.

The second part is equally contentious. In the chorological intervention we identify research themes for a relevant urban studies. There is a long tradition of critical urban scholarship that we do not engage with in this book. Our 'alternate' approach means distinctly different critical research; there is no intention of replacement of other critical urban research – we are simply ploughing a different furrow. Aspiring to a trans-modern scholarship creates research questions that are not generally considered in urban studies. Perforce our research agendas have a large historical component and the chronological intervention is into nomenclature debates over climate change in history and geology. Starting from basics, our concern for anthropogenic climate change leads us to investigation of its origins because determining how it first happened – which we link to urban demand – is crucial to understanding it today. In his work on climate change Moore (2014: 4) contends:

> Conceptualizations of a problem and efforts to resolve that problem are always tightly connected. So too are the ways we think the origins of a problem and how we think through possible solutions.

It is in the spirit of this argument that this chapter has been written.

Building a narrative

We use Jane Jacobs' (1970, 1984, 2000) theory of cities as economic development to integrate the urban with the temporality and spatiality of anthropogenic climate change. Jacobs' writings on economics have had a mixed reception in geography (Jonas, 1986; Jonas and Ward, 2007) but, as reported in Chapter 2, leading economists have been much more accepting of her ideas (Nowlan, 1997) from both radical (Krugman, 1995) and conservative perspectives (Glaeser, 2011). Notwithstanding these debates, we use her urban economics because it is specifically appropriate for our critical ends in three key ways.

First, the theory is articulated on inside/outside relations of cities expressed as export and import replacement modes of economic growth. This process of cities as the engine of economic development integrates economic activities within cities with connections between them. The dominant import replacement activity in this theory is a continuous generation of home-grown production, a localisation creating concentrated, complex city economies within wider urban networks. Second, this conception of city process is set in a supply and demand framework wherein we interpret the latter that is the root cause of change. This will be shown to be very important for understanding anthropogenic climate change in contrast to policy prescriptions that generally engage with supply, most notably, state negotiations over international carbon reduction. Our prioritising cities à la Jacobs switches concern emphatically to demand: rapacious urban demand. Third, and crucially, Jacobs advocates early urban development that approximately coincides with revisions being made to the timing of anthropogenic climate change by William Ruddiman (2010; Taylor et al, 2015), which we use in detail later.

Generics: understanding the material power of cities

Jacobs' (1970, 2000) theory of urban economic development builds on her classic book on US cities (Jacobs, 1961: 442–62) in which she famously concluded that 'organized complexity' was 'the kind of problem a city is'. This promoted bottom-up agency over City Hall. There are hints of this idea in some city and climate change writings, for instance, Bulkeley (2013: 11), referring to a Foresight (2008) report that identifies the 'self-organization of independent actors' as 'radical possibilities for living different urban lives in reconfigured urban economies'. What we aim to do in this section is to tie down such thinking specifically to the nature of cities as the loci of economic development. This was introduced in Chapter 2; here we add detail to our Jacobs' legacy description. The theory is presented in two parts; first, identifying the special nature of cities and city life, and second, delineating the mechanisms that result from this specialness.

The communication advantages of cities

Cities emerged at various times and places across the world as special settlements orchestrating local and non-local relations. According to Jacobs (1970: 35):

> ... in modern and historical times, no creative local economy – which is to say, no city economy – seems to have grown in isolation from other cities. A city does not grow by trading only with a rural hinterland. A city seems always to have implied a group of cities, in trade with one another.

Living in a city as part of a network of cities means a whole new world of enhanced communication, in terms of both quantity and quality. For most of its existence, humanity lived in hunter-gatherer bands of about 150 people involving just intra-band communication plus local contact with neighbouring bands. In contrast, the invention of cities and their networks enabled daily contact with far larger numbers of people, both within and between cities. In addition to this quantitative boost, there was a qualitative effect: regular network relations between cities produces contact with a diversity of non-local people, some of whom move to live in new cities, creating the beginnings of urban cosmopolitanism. It has been estimated that in Catalhörük about 7,000 BC (in Anatolia, one of the earliest known cities) residents had a human communication enhancement greater than pre-urban people by a factor of nearly 2,700, and for Uruk (in Mesopotamia, perhaps the first great city), about 4,000 years later, the enhancement factor was over 550,000 (for details on computation, see Taylor, 2013: 99). In other words, living in cities transformed the density and diversity of communications of people providing direct access to so much more information and knowledge than could ever be experienced in previous non-urban worlds of small human bands. Although measured quantitatively, this is only to provide a sense of the scale of communication difference cities created. In reality this is best viewed as a fundamental qualitative shift: cities offered a completely new interactive world.

What difference does this make? According to Glaeser (2011: 7), all the additional communication enables city residents to become 'smarter' than their less connected country cousins; not because they are personally more intelligent but because of enhanced opportunities: they have more and diverse people to learn from who are also 'smarter' from being in cities. In particular, he argues that 'cities speed innovation by connecting their smart inhabitants to each other' to become very creative places. This is the reason cities have such a strong historical track record for creative change:

'cities' are literally synonymous with 'civilisations', which are classically defined as societies organised through cities.

But being 'smarter' is not just about economic innovations; living in cities can instigate collective changes in ways of living. For instance, in the second half of the 20th century, environmental concerns for the future focused on global population growth featuring predictions of between 12 and 15 billion people. But Ehrlich's (1971) famous 'population bomb' has somewhat fizzled out, with peak population now expected at about 10 billion. This massive reduction of up to a third is largely due to the changed reproductive decisions of women living in cities. As Brand (2010: 55) tells it:

> City dwellers have few children – the billion squatters like everyone else. Thanks to a by-product of urban growth, the core environmentalist panic over population is quietly being undermined.

This example of mass behavioural change – what Pearce (2010: 246) calls 'the feminization of cities' – shows the potential transformative power of cities in our century.

Glaeser (2011) has titled his book *Triumph of the City*, reflecting the oft-quoted fact that urban dwellers are now a majority of the world's population; he asserts that we are now 'an urban species' (2011: 1). But passing the 50 per cent threshold is just part of an incessant trend, a global urban change that is expected to result in perhaps three-quarters of the world's population being urban around mid-century. From the perspective of our communicative-enhancement argument, this change has potentially immense consequences: the 21st century's billions of people are singularly special because, as largely urban dwellers, they encompass city potential for changing ways of living at an historically unprecedented rate. But before we explore the real significance of this, it is necessary to specify more precisely the actual mechanisms that make cities so potent.

Jacobs' mechanism of development

When translated into urban economics the two sources of communication advantage – dense and diverse links – are treated as externalities. An externality denotes the context within which a firm operates that is not market-defined (that is, it is 'external' to the market). Cities provide two important positive externalities for firms: (i) agglomeration/cluster externalities within cities whereby firms can take advantage of being close to other firms, and (ii) network/connectivity externalities between cities whereby firms can take advantage of connections with other cities. In combination these make cities into rich places of information sources and knowledge flows. As knowledge hubs largely based on face-to-face contacts in a learning milieu, cities are where both intense and cosmopolitan economic environments are to be found.

According to Jacobs (1970), economic development is a special case of economic change. It is generated by two master economic processes, innovation and import replacement, which are both features of city creativity. In this argument innovation is a function of the size and complexity of cities, where urban problems generate new demands that only these creative places can satisfy through new productions and consumptions. Import replacing derives from the diffusion of innovations through city networks of creative places, where innovations can be creatively improvised for local productions and consumptions. Both processes generate new work, thereby increasing the complexity of a city's division of labour. It is this dynamic outcome that specifically defines economic development. An economy can grow by just increasing what is already being produced – adding more old work to existing old work such as doubling the output of a factory – but this economic change does not diversify a city's division of labour and hence does not qualify as development. Thus cities are exceptional settlements because of their complex dynamic divisions of labour resulting from their innovations and import replacements: it is these processes, linking concentration and connectivity, that make cities the prime units of economic development.

This theory posits a world divided between two types of settlement with contrasting economies: complex economies of cities mutually linked through networks, and simple economies of small towns, villages and farms dependent on cities. Jacobs (1984) describes in detail how this division enables cities to project their economic power and mould far-off economic landscapes to their own needs. The crucial point is that this is a demand theory of economic change wherein demands generated in cities transform economic activities and fortunes in both urban and rural realms. Put simply, what we argue is that cities are inherently demanding, from which anthropogenic climate change is an unintended consequence.

Linking to Ruddiman's early anthropogenic climate change thesis

A key antithesis (X) from the previous chapter used Ruddiman's (2003, 2010) argument that anthropogenic climate change is much older than modern industrialisation, many millennia older. The evidence supporting his thesis has been thoroughly evaluated in Kaplan et al (2010) and Ruddiman (2013). His basic method is to show how the Earth's climate varies over the long term – what he calls 'nature in control' – and then to identify change anomalies in the recent past, which he argues can only be explained by human activities. In this section, first, we present a summary of Ruddiman's thesis detailing the processes, and second, we bring cities into the argument.

Ruddiman's time lines for early anthropogenic climate change

The thesis for change from 'nature in control' to 'humans in control' can be set out as follows:

1. Nature in control is described for both high latitude ice ages and tropical monsoons in relation to the Earth's orbital changes. These two cases are related to atmospheric concentrations of the two greenhouse gases, carbon dioxide (CO_2) and methane (CH_4) respectively.
2. Glacial cycles were superimposed on a longer-term cooling trend about 2.75 million years ago and are determined by changes in the Earth's orbit that affect the amount of solar radiation received. Cycles of 41,000, 22,000 and 100,000 years are identified in a sequence of between 40 and 50 'ice ages'.
3. Alignment of radiation maxima from orbital tilt (41,000 years) and precession (22,000 years) produces radiation peaks approximately every 100,000 years, removing ice sheets from the northern hemisphere. The most recent occurred from 16,000 years ago, leading to minimum ice cover 6,000 years ago.
4. In terms of CO_2 concentrations in the atmosphere, Ruddiman finds peak values in relation to the 100,000-year cycles: for the three cycles before the most recent, the peaks occur just before the minimum ice cover.
5. In the current cycle the natural CO_2 peak is found 11,000 years ago, before minimum ice, as previously noted, but the subsequent decline of CO_2 does not continue as expected: about 8,000 years ago the CO_2 trend reverses direction, showing increasing concentrations of CO_2. It is this anomaly that Ruddiman interprets as humans beginning to take control of climate change.
6. The strength of monsoons is driven by changes in solar radiation caused by the 22,000-year orbital cycle producing a wet–dry climatic sequence.

7. A summertime maximum for solar radiation in the northern tropics occurred 11,000 years ago, since when radiation levels have declined to a minimum, so that we are halfway through a 22,000-year cycle today.
8. In terms of CH_4 concentrations, these are generally at a maximum at the peak of solar radiation – the wet period produces more vegetation – and CH_4 levels subsequently decline as the solar radiation reduces and a dryer climate prevails.
9. In the current cycle CH_4 levels do decline from 11,000 years ago, but this abruptly stops and reverses 5,000 years ago. From this point the concentration of CH_4 in the atmosphere increases. This second anomaly is interpreted by Ruddiman as a further effect of humans beginning to take control of climate change

In creating this early anthropogenic climate argument, Ruddiman specifies separate processes to account for the two greenhouse gas anomalies at 8,000 and 5,000 years ago, both as consequences of agricultural activities. Increasing CO_2 from about 8,000 years ago results from the transformation of land cover due to large-scale deforestation for farming. Increases in CH_4 from about 5,000 years ago results from the development of wetland rice production in tropical Asia through a dryer period that maintained and increased rotting vegetation levels. In both cases he argues that human-induced land cover change modified terrestrial ecosystems to such a degree as to change the global climate. Thus Ruddiman's (2010: 6) seminal contribution to climate change science is to add an initial slow millennial rise in greenhouse gases to the conventional identification of a rapid rise in greenhouse gases over the last 200 years. However, evoking the first stages of agricultural development as the human activity generating early terrestrial atmospheric alterations has received particular scepticism within climate science. And this is where Jacobs (1970) re-enters our argument by linking cities to agricultural revolutions in her 'cities first' thesis (antithesis IV). And making this link creates a contribution to Ruddiman's defence of his early anthropogenic climate change thesis by providing a much more powerful social mechanism of change.

Bringing cities in

A very obvious response to the early anthropogenic climate change thesis is that there were simply too few people living in these early times to have made such an impact. Ruddiman has countered this criticism in three main ways. First is the time effect: the impact of changing land cover has been compounded over a very long period. Second, the numbers of people involved in the rise of agriculture have been widely under-estimated; in contrast, Gignoux et al (2011) show population growths in double digits across different regions when agriculture appeared. Third, Boserup's (1965) thesis on the relationship between agricultural intensification and population growth is used to indicate that because the earliest agriculture had less population pressure, it initially produced very high per capita land cover changes (Ruddiman and Ellis, 2009; Kaplan et al, 2010). These arguments are brought together in Ruddiman's (2013) state-of-the-art summary of his thesis, and we have no disagreement with them. Rather, by bringing in Jacobs' cities first thesis we add a fourth counter to the idea that early people could not have affected global climate change. Not only were there more people than generally thought; they were also beginning to be settled in a new geography – cities. But bringing in cities is not just an additional support to Ruddiman's thesis; it provides an explanation for why and how agriculture begins, and it produces a common mechanism of social change capable of creating climate change for both of Ruddiman's slow impact era (early urbanisation) and rapid impact era (modern urbanisation).

The key connection between Jacobs' and Ruddiman's ideas is that for the former agriculture is a very significant import replacement for the earliest cities (antitheses IV and VI from the last chapter). Embryonic cities derive from Neolithic trade networks when production (new work) is added to trading camps. With the consequent urban population growth, existing food procurement from hunter-gathering practices is unable to satisfy the increased demand: agriculture is invented to meet the shortfall. This is Jacobs' (1970) 'cities first' thesis: cities emerge in the human story before farming. The conventional view of the very first cities appearing in Mesopotamia several millennia after the coming of agriculture is based on a supply theory of urban origins – cities only appear when agriculture has advanced enough to generate food surpluses to support urban life. To counter this traditional view the alternative demand theory simply asks why hunter-gatherers themselves – remember they are Sahlins' (2004) 'first affluent society' – should choose to invent agriculture and the concomitant extra work (Taylor, 2012a, 2013: 102–13). Jacobs' answer is that there was no such rustic invention; agriculture is an early expression of the power of cities to mould their environs for their specific needs. Evidence for early cities across the world is briefly reviewed in Taylor (2013: 138–44). From an orthodox archaeological perspective these large settlements in 'wrong places at wrong times' cannot be locales for inventing agriculture because they are not recognised as cities; this issue has recently been debated (see Smith et al, 2014; Taylor, 2015). However, Jacobs' shift in interpretation of farming origins in no way lessens the importance of agriculture in terms of the changing land cover outcome, as argued by Ruddiman; rather, it interprets agriculture as an intermediate step in a process that has its origins in urban demand.

The point we are developing here is not just the idea of farming not happening without urban demand, but that it is through understanding the city economic development mechanism that we can know the fundamental social practices behind human-induced climate change, and which will apply to both its slow and rapid phases. Thus cities are the crucibles of world-changing transformation like agriculture, and this is why modern industrialisation is associated with historically unprecedented levels of urbanisation. Drawing on Jacobs' cities first thesis, Soja (2000) identifies three urban revolutions: the first about 8,000 years ago, comprising the origins of urbanisation represented by Jericho and Catalhörük; the second about 5,000 years ago is the familiar Mesopotamian urban revolution (traditionally 'the first cities'); and the third is the massive urbanisation of the last 200 years associated with industrialisation. The temporal matching with Ruddiman's thesis is remarkable (Taylor et al, 2015): Table 4.1 provides the parallel details. Of course, temporal matching does not equal causality; for this there has to be a process to explain the coincidences. In this case it is our urban demand argument. Put these together and we can produce an original geohistorical narrative both for understanding anthropogenic climate change and for how we should respond to it policy-wise.

Table 4.1: Soja's and Ruddiman's time lines

Time	The 'anthropo bit': Social science Soja's 'urban revolutions'	The climate change bit': Physical science Ruddiman's 'human control of climate'
10,000 years ago	The *first urban revolution*: widespread land clearances for food	Rise in *carbon* in the atmosphere; slow rise in global temperature contra to natural climate cycles
5,000 years ago	The *second urban revolution*: the rise of riverine civilisations of cities	Rise in *methane* in the atmosphere; slow rise in global temperature contra to natural climate cycles
Modern	The *third urban revolution*: the rise of the capitalist world economy	Industrialisation; *rapid rise* in global temperature

A trans-modern narrative on anthropogenic climate change

Humans are different from other species in having two means of social reproduction. All species harness the immediate resources of their environment, thus creating a local dependence. The local can be a fixed territory or a moving supply area (seasonal path). Uniquely humans also draw on non-local resources, movement of goods to them from afar (Jacobs, 1992). This has always been very clear from archaeological studies, where non-local artefacts – stone tools from distant specific geological sources – are present in excavations. Archaeologists call this 'the release from proximity' (Rodseth et al, 1991: 240); although Gamble (2007: 211) has noted that material from further than 'daily foraging range' usually constitutes less than 1 per cent of an excavation assemblage, it does represent 'the local rule broken', and that is the key point. It is not the quantity that matters; a qualitative difference in the manner of social reproduction has been broached. However, the problems of developing the necessary trust with strangers to enable routine trading kept the non-local contribution to social reproduction quantitatively unimportant (Curtin, 1984; Graeber, 2011). Overcoming this problem enabled the rise of cities.

Evidence of the earliest agriculture is about 12,000 years ago, but Ruddiman (2010) shows its climatic impact from only 8,000 years ago. This time lag represents a response time of a slow transition, the first stirrings of urbanisation, a stuttering into being of city networks with their new demands for food (Soja, 2010). These initial networks are relatively fragile; overall growth is slow, but by 8,000 years ago there is clear evidence of substantial urban settlements trading through complex divisions of labour, as evidenced by Jericho and Catalhörük (Jacobs, 1970; Soja, 2000, 2010; Taylor, 2012a, 2013). This is the breakthrough of the non-local becoming significant in macro-social change. It is this development of multiple vibrant city networks whose creation of new landscapes of agriculture instigates the turnabout of CO_2 that Ruddiman (2010) reports.

However, the long-term resilience of early cities was initially problematic: surrounding agricultural supply became more distant as soils were exhausted, finally creating empty quarters and lost cities (Pauketat, 2004; Taylor, 2013). This outcome is overcome 5,000 years ago, with the Mesopotamian city network based on sustainable irrigation agriculture. Ruddiman (2010) emphasises tropical Asia's rice cultivation at this point, which he links to rapid population growth; here this is interpreted as early Chinese city networks creating large new demands for food. South Asian and Egyptian riverine cities can also be pushed back to 5,000 years ago. Collectively this is the beginning of the era of large cities in large networks of cities. The first city of 40,000 people is identified in 3300 BC as Uruk in Mesopotamia; by 2800 the city has a population of 80,000, and with 10 other cities in its network, the urban population of the region is over a quarter of a million (Modelski, 2003: 22; Taylor,

2013: 114). Beyond mere demographics there were behavioural changes enhancing consumption. Leick (2001: 10) refers to 'the importance of collective feasting, sometimes called 'conspicuous commensuality' – a precursor to 'conspicuous consumption'. Overall, this new level of urban demand requires a social reproduction logistics at a whole new scale, not to mention the outputs, both exports and waste (human and animal). It is the rise of large cities and their consumption and disposal effects that instigates the turnabout of CH_4 that Ruddiman (2010) reports.

Importantly, such new scales of social activity require new means of governance: the world-changing invention of states (Smith, 2003; Yoffee, 2005), first as city-states and then empire-states incorporating multiple cities (Taylor, 2012a, 2013). This new political dimension is vital to understanding the relatively slow rise of large cities before the modern era. What is happening is that in traditional empires the city development process is replaced by plunder and tribute as the main means of wealth accumulation (Taylor, 2013). The largest cities are now typically the imperial capitals that have become huge urban centres of consumption, but otherwise there is a societal constraint on city development as new work: in all tribute empires, large and small, the urban population is typically less than 10 per cent of the total population (Taylor, 2013). However, both empires and cities continued to grow, and therefore overall the trend is a gradual increase in the collective large urban population (Taagepera, 1978; Chase-Dunn and Manning, 2002), which keeps Ruddiman's (2010) greenhouse gas anomalies moving forward.

In this period of world empires the rise in CO_2 ceases rising about 2,000 years ago, which Ruddiman (2010: 87) explains by the Boserup effect: we would describe this as being the result of increased urban demand leading to improved agricultural technology, and thus a reduction in per capita land cover needs. Apart from this clear change in trend, there are also what Ruddiman (2010: 119) calls 'CO_2 wiggles', small variations in the overall pattern. He explains these by reference to large-scale disease epidemics and pandemics that, by decreasing the population, lead to less agriculture and therefore cleared land returning to being covered (2010: 132–3). However, this can be interpreted as a clear case of fluctuations in urban demand: these diseases are disseminated through city networks and their effects multiplied in dense urban populations (Verbruggen et al, 2014: 50–1; Hassett, 2017: 209–28).

Traditional world-empires begin to be superseded by the modern world-system in a transition from about 1450 to 1650 (Wallerstein, 1974, 2004). A different historical system emerged where the balance between political and economic elites is readjusted in the latter's favour to create a more balanced power relation between cities and states (Taylor, 2013).

The result is a new release of urban economic development potential. However, at first the effect of the new social relations is not reflected in global urban demand increasing CO_2 levels. The obvious reason is that the modern world-system was initially not global, being focused largely in Europe and the Americas. But there is a clear effect on climate change by this new historical system. Initially, from 1600 to 1800, CO_2 levels actually fell in one of the larger 'wiggles' (Ruddiman, 2010: 87). So where is the reduced urban demand to cause this decline? An explanation can be found in the greatest disease pandemic of all, which afflicted the Americas after European contact (Ruddiman 2010: 132–3). This decimated indigenous agriculture and the cities consuming the production, both of which have been conventionally under-estimated – Mann (2011) provides an overview of this demographic process, while evidence for urban Amazonia is set out in detail by Clement et al (2015). More generally in European imperialisms, temperate biomass was reduced before the denser tropical biomass, leading to later CO_2 and CH_4 rises (Crosby, 2004). Thus the great greenhouse gas increases precipitated by the modern world-system have an important and very clear time lag.

Ruddiman (2010: 171) starts fast anthropogenic climate change conventionally in 1800, and we can link this change symbolically to our argument through the leading cities of the period: in 1800 Beijing, capital of a traditional empire, was still larger than London, but very soon London became the world's largest city through most of the 19th century and beyond. In this time China became incorporated into the modern world-system; the latter became fully global by 1900. This is shown in the enormous increase in large city populations, indicating a truly qualitative shift to a new world of great cities as recognised at the time (Weber, 1899). Thus, instead of labelling the social change as simply 'industrial', the cities approach reinforces Brooke's (2014: 480) argument that the key environmental trigger is immensely heightened urban demand becoming worldwide. For instance, the great urban expansion includes explosive city growth in the frontiers settled by English-speaking people, notably Chicago first, followed by Melbourne, as described by Belich (2009) (see also Taylor et al, 2010).

However, the urban change in the 19th century is more than simply demographic. New ways of living are created. In the modern world, industrial cities changed traditional ways of living into modern clock-based disciplined behaviour in places (factories) and flows (railways). And most important for our argument here, in 19th-century cities in major countries around the world, traditional bourgeois thriftiness morphs into consumption-driven behaviour (Laermans, 1993; Dauvernge, 2008) as a new way of living (symbolised by the invention and diffusion of the departmental store from the mid-19th century). This behaviour has been inherited and exported as continuously expanding global demand (symbolised by the invention and diffusion of the shopping mall from the mid-20th century). This fundamental change in bourgeois behaviour was the consequence of a cultural transformation of commerce from being just one element of traditional societies to becoming the dominant feature of modernity as capitalism (Wallerstein, 2004: 23–41). This social change finally releases unremitting city economic processes, resulting in spirals of ceaseless capital accumulation requiring ever-expanding consumption as a truly mass demand.

Thus our narrative ends by returning to how it began, the unique human process of social reproduction. It is the ecological implications of mass-demanding cities that are toxic. Harnessing non-local sources of reproduction enabled by trading has today culminated in what Jacobs (2000: 119) calls a 'cultural breaching' of the behaviour necessary to ensure ecological sustainability. She argues that reproduction of any species requires 'traits which prevent it from destroying its own habitat'. For instance, many large animal species develop behaviours that check habitat destruction – these occupy time that could otherwise be devoted to material reproduction. Seemingly 'social activities' prevent damage to the environment that would result from continuous grazing by, for instance, elephants, or incessant hunting by, for instance, lions. It is this 'social' behaviour that prevents full-time exploitation of an environment to its destruction and thereby facilitates reproduction of the herd or pride. Humans have, of course, highly developed social behaviours that have checked habitat destruction for most of their existence through reproduction as hunter-gatherers, the leisure-rich 'first affluent society' as described by Sahlins (2004). But the danger enhanced non-local reproduction poses to natural 'dynamic stability' is a very different matter, and a few millennia of cities and a few centuries of capitalism have finally undermined whatever evolutionary safeguards were in place.

How does this narrative end? We don't know, but things are not looking good – that is the message of the climate emergency. Since this is a story of cities it follows that cities will be at the heart of the narrative's denouement, whatever form this takes. Navigating this difficult terrain is the subject of the final chapter.

Trans-modern interventions

Having a new narrative enables us to revisit the time and space framings that were introduced in Chapter 3. Familiar modern chorology and chronology can now be replaced by trans-modern ordering of humans in nature.

New chorography: geographical imagination

We have re-positioned how we think about the relations between cities and anthropogenic climate change. This has two important implications. First, our trans-modern thinking opens up myriad avenues of potential novel research enquiry. This entails both fresh looks at old agendas and building new agendas. Second, the focus on cities brings spatial organisation to the fore. This brings geographical imagination into play as a necessary research input. Here we identify five crucial areas of study for developing critical research agendas in urban studies: (i) adding cities to the 'Neolithic package'; (ii) revising historical urban footprints before industrialisation; (iii) policy and the cities-as-demand mechanism; (iv) framing a politics of change through cities; and (v) locating cities in a social theory of reproduction. We provide a skeletal outline of agenda building in each research area through a focus on the key issues involved.

Our approach brings the question of urban origins back to the fore. As Angus (2015: 1) argues, 'determining when radical physical changes in the Earth system happened provides a basis for determining which human activities were responsible, and thus what measures humans might take to prevent the change from reaching catastrophic proportions'. But historical urban geography research has not been much concerned for this question since Harvey's (1973) discussion of Wheatley (1969), although slightly later Carter (1983) did provide a review of ongoing debates including Jacobs' (1970) cities first thesis. More broadly and recently, urban researchers have left the matter to archaeologists, specifically meaning endorsement of Childe's (1950) theory of urban revolution wherein the first cities were built in Mesopotamia 5,000 years ago (Le Gates and Stout, 2016: 30–8). Superseding this traditional supply theory of urban origins remains controversial (Smith et al, 2014; Taylor, 2015) and yet, is central to our extension of Jacobs' ideas into anthropogenic climate change. A new research agenda arises from the historical demographic research that Ruddiman deploys on challenging conventional climate change thinking.

This research agenda is about adding cities to the 'Neolithic package' of agriculture, villages, pottery and other clay or stone objects. As previously mentioned, Ruddiman's early anthropogenic climate change thesis requires a reassessment of Neolithic populations clearing land cover through agro-pastoral practices. Lemmen (2009) has used simulation models combining population totals with land cover maps to estimate induced carbon releases that are too small to support the changes in carbon reported by Ruddiman (2003). Ruddiman (2010) has countered this using larger demographic estimates and higher land clearance per capita ratios; in new modelling (Kaplan et al, 2010), Ruddiman and Lemmen are reconciled on this. But this should not be the end of the matter. To estimate land cover removal required input for population reproduction: Lemmen uses Gregg's (1988) model of a Neolithic village of 30 people. This means simulations are based on subsistence farming in a landscape scattered with multiple very small villages. Without an urban dimension, this betrays an extreme poverty of geographical imagination.

Jacobs' (1970) cities first thesis envisages a much more complex landscape of networks of cities requiring 'rural' land clearances for food, fuel, construction and other uses. Adding this urban geography to the simple demographics of existing simulations is a research agenda that both engages with the initial debate over Jacobs' work as well as informing the Ruddiman climate debate. Data is available from earlier simulations; the new research involves adding cities and resulting complex divisions of labour to create new estimates of induced carbon emissions from both clearance and wider uses of fire. In the original simulations there was just 20km^2 of forest required for every hundred people practising subsistence agriculture; in an urban landscape with farmers responding to urban demand there is obviously a need for a much larger forest resource and clearance. By how much? How can you estimate this? What effect does this have on other parts of the simulations? These are the research questions for urban geographers to explore.

Beyond urban origins, there is a general disregarding of pre-industrial urban populations. This is not for want of data: historical demographers provide population estimates for large cities for the last 5,000 years (Chandler, 1987; Modelski, 2003). But even when these are employed, the poverty of geographical imagination persists. Nelson (2014) provides a regional analysis of world population that includes global urban population estimates mentioning the leading cities for each period he studies. But an emphasis on just the largest cities rather than complete regional networks of cities means vast numbers of urban dwellers are simply missed out. For instance, some preliminary work focuses on 300 BC, the first period when we have population estimates for more than 50 cities. But these known cities are just the tip of an urban iceberg. Nelson (2014: 36) provides an urban population estimate of 2 million; filling in the urban networks for lower levels in the six regions with cities in 300 BC suggests a figure of 14 million urban dwellers (Taylor, 2017b). Such a seven-fold quantitative difference indicates a basic qualitative difference in how early societies are imagined. It underlines a need to critically assess the modern mindset when dealing with urban matters outside modernity. A critical urban studies can totally revise urban footprints before industrialisation; even millennia ago, concentrations of 14 million people represent a huge urban demand with consequent huge environmental effects.

The first two research agendas are necessary to consolidate evidence for Ruddiman's long, slow anthropologic climate change through urban process and to show how the cities-demand mechanism may have initially operated. Our narrative rests on this mechanism, which itself constitutes a second broader research agenda. It challenges the prime policy emphasis in climate change: national planning for carbon reduction. In UN conferences such as at Paris in December 2015, states negotiate under conditions of severe lobbying by non-governmental organisations (NGOs) to produce targets and timetables to cut their carbon contributions to climate change. As argued in the last chapter this prioritising of supply over demand has serious dysfunctional implications. It is only partially corrected by UN-sponsored local initiatives; it never recognises cities as innately demanding.

That cities are inherently demanding, meaning that they grow and prosper through expanding consumption by urban dwellers, is not a major theme in urban studies. In fact, this key Jacobs' insight is largely implicit at best, perhaps because of a tendency to avoid conceptualising the contemporary urban in basic economic terms. Thus in describing the 'operational landscapes' of 'planetary urbanisation', the relationship between the conventionally urban and 'the urban formerly conceptualised as rural' is portrayed in terms of the latter 'supporting' the former; the 'entire planet is being marshalled 'to serve … urban industrial development' (Brenner, 2014: 20–1). This tilts research towards supply patterns of infrastructures, and potentially neglects the cities as the demand complement to all this supplying. To be sure cities have egregious effects worldwide, but the whole story begins in the cities. They do not just 'hinge upon and transform operational landscapes' (Urban Theory Lab-GSD, 2014: 474); they are the process of demand that produces the operational landscapes. And this is the central point of our approach to anthropogenic climate change. It requires extensive construction of urban geographies of growing global demand. A simple starting point might be to reconsider von Thünen-style urban land use models with their circular bands of decreasing densities. Today's outer ring comprises hectares of car-parking spaces: a serious research project on the geography of car parking at both city and global scale is long overdue. In any future reinvention of the city this particular geography will represent the path dependency in physical form confronting radical change.

But the key point from our narrative is that modern mass consumption developed in 19th-century cities was a bottom-up process of acquisitive behaviour; embedded into modern life, to reverse the now mass-acquisitiveness will also be a bottom-up process. Since this development was not created by political (that is, top-down) policy-making – global, national or local – we should not assume it will be 'solved' by the actions of today's politicians – presidents, prime ministers or mayors – or their policy-makers, whatever their political will. This is not to say that formal policy-making and implementation is irrelevant (for example, on reducing carbon emissions), but states negotiating in UN conferences over supply issues while the everyday behaviours of billions of people keep the demand ever increasing is simply bizarre. We agree with Pieterse (2008) that city futures depend on 'everyday urbanism', the continual remaking of cities by their residents and workers. Further, and in light of the urgency in climate change policy, our approach goes beyond intensive research to understand contemporary consumptions. In order not to be part of the current inadequate top-down process, critical urban researchers need to address their methodologies in both theory and practice. A trans-modern approach points towards specific methodologies such as participatory action research (Askins and Pain, 2011) where people are not the subjects of the research but become 'producers of knowledge in the research process' (2011: 806).

This brings us to the politics of climate change through cities. The anti-globalisation movements interrupting global governance meetings and the global Social Forums initiated at Porto Allegro and their ilk are necessary political challenges to a dysfunctional world (Smith et al, 2011), but in our account they are an insufficient, modern reaction. Collective radical pressure culminating in national revolutions (that is, as political events) has been successful in the modern world, but we are concerned with a 'long revolution', a transition beyond modern consumption. This is a trans-modern shift to change the everyday actual behaviours of billions of people, and, just as important, to change the 'development' aspirations of other billions of people to realise such behaviour. This is the challenge of mass consumption, and it is essentially an urban problem. But what political urban studies does this imply?

Our interpretation of the problem is that the economic process in cities has colonised the social process as modern city living so that shopping, in its ever-wider range of guises, has become the dominant leisure pursuit. From a trans-modern approach we know that this does not have to be how cities grow: for instance, historically urban guilds were as much social (religious) as economic (trades). But such a 'historical lesson' is not a marker for the future; this is an area of study that requires some utopian thinking. Current research on green/sustainable cities is proximately relevant, but again betrays very limited geographical imagination: a world of green cities scattered across the world's urban landscape is no more insightful than a supposed Neolithic scattering of little villages. There can be no meaningful politics of change without a vision of what might be achieved, and in this case, a planetary urban vision, a future global urban studies, is essential for reconstituting a 'new' social from its immense domination by material consumption.

We are suggesting a critical urban studies research agenda combining one key aspect of Jacobs' city growth mechanism with Harvey's (2012) recent harvesting of Bookchin's (1995) urban utopianism, but in a broader urban landscape that is now global. As we have shown, Jacobs' (1970) theory of urban economic development is essentially a process of continuous localisation of innovative change (import replacement), but she also makes clear that not all innovation is economic. Cultural, social and particularly political innovations abound in cities along with economic innovations; whether formal or informal, ethical or non-ethical, legal or criminal, they all make up the maelstrom that is urbanism. Currently, research agendas on cities' agglomeration and connectivity practices are largely in the disciplinary territory of economics/business studies/economic geography; by emphasising their broader urban substance we can explore a trans-modern politics that bypasses modern state-scale political parties as instruments of change (only invented two centuries ago) and search out newly invented political mechanisms conjoining the urban everyday with the urban global. This combines the right to make the city with the right to make the world. The planetary urbanisation initiative (Brenner, 2014) is working towards this end; our approach suggests bringing anthropogenic climate change to the centre stage of their emerging agenda as a political urban geography, where blue-sky thinking is encouraged in the light of planetary urbanisation being 8,000 years old.

The final agenda derives from understanding that our approach implies much more than a historical reframing of urban geography research. Over several decades, urban research agendas in geography have been strongly influenced by David Harvey's original Marxist exegesis, wherein city processes have loomed large (Harvey, 1973, 1982/1999, 1985). In this work the urban in multifarious ways is consistently treated as an outcome of broader social processes. Allen Scott (2012; Scott and Storper, 2015) is also explicit on this: first know your society and then study its cities. These approaches do not mean that cities are not important, for example they have had vital roles in overcoming financial crises, but they are not the crucial mechanism undergirding social reproduction. But as outcomes they are designated of secondary importance. Our critical approach turns this around: cities are the instigators of macro social change, not its outcome. This is the most overt way that a trans-modern approach challenges modern urban scholarship. We are suggesting that the latter, in both its traditional and radical forms, may have been useful to creators of knowledge for modern purposes, but this is no longer the case for a world facing a global climate emergency.

Nevertheless we are witnessing a convergence of thinking in this area. First and most crucially, in recent years Harvey (2014) has moved from an emphasis on urban as product to urban as process drawing on the analysis of circulation of money, production and commodity in the second volume of *Das Kapital*. This matches our emphasis on city growth as demanding. Second, for Harvey the radical political agenda must be played out on a global and local basis, in urban locales rather than nation-states, with rebel cities generating political innovations (Harvey, 2012, 2015). This clearly parallels our thinking. Third, at this juncture of capitalism there is a need to embrace a myriad of radical groups, including anarchists, rather than a monolithic single national party to provide opposition to capital. Here we see a huge critical urban studies agenda on the myriad of activist groups making and dreaming of how a safe urban world can be produced for everyone worldwide. We broach the relations of climate change politics to other radical politics in the final chapter. But ultimately we need to devise research into the Advertising-Big Data-Social Media complex that controls the future of global urbanisation. This is a second key challenge to the politics of anthropogenic climate change that we deal with in the final chapter.

Last, but not least, our agendas require the long overdue integration of research on 'Global South' and 'Global North' cities, which have tended to go their separate ways (Robinson et al, 2016). A 'global' geographical imagination of cities is a key unintended benefit of bringing anthropogenic climate change to centre stage.

New chronology: Urbanocene

The purpose of any new narrative is to influence both research and politics as previously discussed. And one fundamental way of achieving this is to challenge existing language about how a subject matter is described. This can be as basic as naming what is happening. We have provided a new narrative on anthropogenic climate change and we conclude the discussion by intervening in an ongoing debate about what to call our current times. As well as familiar historical periods, there is a science of naming different times in the evolution of the Earth. It is within the latter's organisation of time that discussions are to be found on how to accommodate anthropogenic climate change into the geological timetable, and so this is where we intrude.

In the early 20th century Milutin Milanković, in solving the mystery of ice ages, understood that the varying pathways of the Earth in the solar system were matched by, have led to, different environmental eras. This, combined with the tectonic movements of the Earth's crust, have produced different geologic time periods. The resulting geological timescale is used by geologists, paleontologists and other Earth scientists to describe the phasing and relationships between events that have occurred throughout Earth's history. This naming of eras is necessary for coordinating research in different fields through characterising different periods in terms of the huge variety of events, for example cold and warm periods, explosions of life and mass extinctions. In recent years such thinking has added a new era to the environmental lexicon: the Anthropocene, a world defined by one species – us. Although much disputed by geologists, empirical indictment of humans for both global warming and declining biodiversity has resulted in this remarkable thesis. The use of geological language to denote contemporary environmental changes is very important because it marks a huge shift in human–environmental relations, both actual and in scientific understanding.

The first serious attempts to formulate a geologic timescale that could be applied anywhere on Earth were made in the late 18th century. The most influential of those early attempts (championed by Abraham Gottlob Werner, among others) divided the rocks of the Earth's crust into four types: primary, secondary, tertiary and quaternary. Each type of rock, according to the theory, formed during a specific period in the Earth's history. It was thus possible to speak of a 'Tertiary period' as well as of 'tertiary rocks'. Indeed, 'Tertiary' (now Paleogene and Neogene) remained in use as the name of a geological period well into the 20th century and 'Quaternary' remains in formal use as the name of the latest period starting about two and a half million years ago. Since 1974 the International Commission on Stratigraphy, a sub-committee of the International Union of Geological Sciences, has been tasked with precisely defining the units of time (periods, epochs and ages) of the International Geologic Time Scale, thus setting the global standard for the history of the Earth. We now live in an epoch that is termed the Holocene at the very end of the Quaternary period. This era commenced after the last glacial period some 12,000 years ago, and is characterised by a relatively stable and warm climate. Although human life evolved prior to the Holocene, it is clear that population growth and environmental impacts have grown considerably during this period, along with the development of major civilisations, written history and overall significant transition toward urban living in the present (Fairbridge and Agenbroad, 2018): if the Anthropocene exists, it is part of the Holocene.

Identifying the Anthropocene should be the work of the International Commission on Stratigraphy, but therein lies a problem. By its very name this body of scientists is concerned with layers of rock that were laid down in past circumstances, but the Anthropocene is very recent geologically, with no time to have created such a strata. In fact, there is a quaint irony here: our times can be characterised by great removals of strata (containing carbons), not the laying of new strata! But stratigraphic efforts have been made by imaging geologists in the far future trying to identify rocks from our time. What marker could they use? A popular answer is to use radioactivity in the rocks emanating from the first nuclear test in 1945. The future geological imprint of plastics and concrete have also been mentioned but, looking at this practically from the viewpoint of a future stratigrapher, we might suggest the overwhelming empirical evidence for our times would be many trillions of chicken bones. Hence Gallusocene (the scientific name for chickens is *Gallus gallus domesticus*).

Of course, this is all very fanciful based on a very optimistic view of our species' longevity. With all due respect to the geologists, in reality, naming an 'anthropo' concept must always be a social science matter (Ellis, 2016; Dalby, 2018). Hence, the common date set for this change is the beginning of the 'Industrial Revolution' because this is the time when human activities are thought to have started having a noticeable influence on the global environment (Steffen et al, 2011; Lewis and Maslin, 2015). Some have been quite precise about the industrial date. The inventor of the Anthropocene term, Paul Crutzen (2002), for example, points to 1784. This was the year that James Watt is credited for inventing the steam engine. As we are writing, the International Commission on Stratigraphy has yet to officially approve the term as a recognised subdivision of geological time although, not surprisingly, given their long-term horizons, there is some momentum for the nuclear option because of its million-year longevity and clarity.

We take a historical social science approach to understanding the Anthropocene. The most well-known example of this way of thinking is by Jason Moore (2015) in his studies linking environmental changes to Wallerstein's (1974) identification of a modern world-system developing in the 'long 16th century' (ca 1450–1650). We have previously discussed this displacement of the Industrial Revolution as the origin of the modern world, and Moore takes this argument on board. The modern world-system operates as a capitalist world-economy, initially a trans-Atlantic economy. This social system based on ceaseless capital accumulation, wherein economic growth is a necessity with no limits, defines our current environmental predicament. Hence he renames our times the Capitalocene (Moore, 2015). He presents a very strong case, but not, we think, a compelling one.

Our narrative has incorporated the modern world-system as part of a much longer perspective incorporating other pre-modern world-systems that Wallerstein (1979) refers to as world-empires. These are the great past civilisations that are identified and named for their great cities. We have interpreted these as places of previously unparalleled urban demand. And even before around 3000 BC, there were the beginnings of urbanisation whose dynamism created agriculture and widespread land clearances. Although not using geological carbon sources, the deforestation involving fire was a carbon source of early climate change. By linking this to Ruddiman's revised chronology of climate change, in this chapter we have brought urban demand to the forefront historically in anthropogenic climate change. Thus we argue that because humans existed for over 100,000 years before instigating climate change, Anthropocene is not strictly the correct term. Also, because humans instigated climate change before the rise of capitalism, Capitalocene is not strictly the correct term. Following our narrative, we propose Urbanocene as the most appropriate term for describing the era when humans, in Ruddiman's words, 'control climate'. Note that all the previous suggestions for starting this new era are alternative modern dates or periods (a modern self-centredness our unthinking has targeted). Our promotion of a much earlier date is suitably trans-modern: the Urbanocene.

Our proposed nomenclature has three implications. First, since the urban/agriculture process is currently dated at about 10,000 years ago, there is very little left of the Holocene that is not Urbanocene. Furthermore, the tendency is for this timing to be pushed back further, as more evidence is found. Thus rather than the Urbanocene being part of the Holocene, it is effectively the Holocene. Second, this fits perfectly with Ruddiman's argument that the lack of a new ice age within the last 11,000 years, and hence the Holocene as a distinct geological time, is a direct human effect: the Holocene's defining relatively stable and warm environment is itself a product of urban demand. Third, by naming our times the Urbanocene, we steer attention to where action is needed – contra Ruddiman, we have made our climate, but we don't 'control' it. That is the trans-modern dilemma.

5

Action: Can We Stop Terminal Consumption?

Introduction: I wouldn't start from here...

Growing urban demand, millennia old with modern rapid acceleration, this is what we have identified as the integral social component, the critical 'anthropo', in anthropogenic climate change. Thus the Urbanocene goes back thousands of years to the beginnings of cities, their demand for food and consequent land clearances. This much longer time frame underscores global urban demand as also integral to what our species has long been, what we are now, and what we are still becoming. Urban demand is not the economist's simple demand variable; it is a way of life, a multifaceted social complex.

> So how do you change a worldview, an unquestioned ideology? Part of it involves choosing the right early policy battles – game-changing ones that don't merely aim to change laws but change patterns of thought. (Klein, 2014: 461)

We have entered Naomi Klein's (2014) world of policy dilemmas pitting embedded worldviews against the need to change how we think. Put bluntly, we seem to be locked into a position where we are collectively pursuing a path with terminal consumption as the outcome. Thus the Urbanocene reveals our 'shared DNA' as the Earth's creative, destructive species: our consuming involves both inventively using Nature (including other humans) to satisfy our demand for stuff while simultaneously destroying Nature to produce said stuff. To continue the DNA analogy, it follows that some sort of critical 'gene editing' will be required to make manifest changes. As an 'urban species' we have to reimagine ourselves, that is to say, we must reinvent the city.

To say 'action is needed' is an understatement of immense proportions. And there are myriad practices and policies aimed at mitigating and adapting to climate change, with many more in the pipeline, working in countless places across the world. They should all be supported – every little helps – but they can never be enough unless über urban demand is confronted head on. We can equate our situation with the common urban myth about a lost cosmopolitan traveller asking a local yokel how to reach a destination. The answer given is totally useless: 'I wouldn't start from here'. What makes this amusing is that the conventional knowledge statuses of the pair are turned upside down: local trumps cosmopolitan, the latter confirmed in an unfamiliar dependent role. But the underlying premise to the story is even more interesting: there is a way to the destination, but this knowledge is not available: a map, depicting the bigger picture, is required. In previous chapters we have attempted to make a 'big picture intervention' by building a radical new knowledge for understanding anthropogenic climate change; a focus on urban demand may provide a fresh start on the journey of humans in their relations with the rest of Nature. The ultimate intention is to point towards radical new practices and policies, and this is what we explore in this final chapter.

But we must deliver a spoiler alert immediately – the clue is in saying 'we explore'. The Declarations with which we began this book are shouting out the innate difficulties of moving forward and confronting the climate change emergency. Of course, this is the practice or policy 'here' we wouldn't want to start from. At the current global level, we have interpreted the practice to be as follows. First, a collective of leading scientists pronounces on the need for policy changes that are becoming increasingly urgent. These are the physical – 'hard' – science experts: '"The science" has spoken'. Second, there is an amazingly naïve social science: it appears to be premised on the idea that presentation of the relevant scientific knowledge will eventually result, after careful consideration, in essential evidence-based rational policies and changes in everyday practices. This is not simply a matter of the neglect of actual social science knowledge – it is not a call for '"The social science" also needs to speak'. The latter is innately 'soft': disordered, diverse and contested, reflecting a messy human world of conflicting cultural, economic, political and social interests combined in innumerable ways. There are social science experts, of course, but it is immeasurably harder to bring their expertise together compared to that of the physical scientists. It is true that the relentless push of urban demand has created a more uniform human experience through a growing urban majority of humanity in the 21st century, but this also creates new differences and interests. Any move from the modern focus on the present (consumption) to an un-modern focus on the long term (stewarding) is very hard to even envisage.

So, yes, action is needed, but here we explore the global context of the present and what this can tell us about how we might advance towards Earth stewardship. This requires a critical geographical imagination, moving away from particular concerns for political or economic or cultural activities and trying to see the overall picture, a synthesis of many pursuits in worldwide framing. The argument is that action for combating climate change has to be universal – as humans we are literally all in this together – but there is a crucial geography to where the action can be most effective. Action taken where current reproduction of urban demand harming the Earth environment is especially vibrant, where change is already colossal, must be the focus of worldwide efforts for positive change. This argument is presented in three parts. First we review our knowledge of urban demand: who has it, what they are doing with it and how it might be different. Second, we focus on China, where urban demand is being developed on a historically unprecedented scale: the mechanisms of this specific reality are interrogated. Third, we conclude with our view of what action might be needed to reinvent the city as stewardship. Perforce we move into utopian scenarios – how it might likely come about – possibly through Chinese leadership.

The enigma of urban demand

Economic process, from 'stone age' to modern, is about interactions of supply and demand. However, we have seen that the economics that permeates the IPCC's knowledge input to climate change negotiations is very biased towards supply, and this is directly reflected in policy focusing more on supply (carbon) than demand (stuff). Nevertheless, beyond this particular discourse, different emphases in policy reflect different schools of economic thought. Keynesianism is especially associated with foregrounding demand, but this is actually about managing consumption to enable supply (production, economic growth) to recover. And this is demand at the national level; our approach foregrounds urban economies, relegating so-called 'national economies' to political jurisdictions of economic processes (Taylor and Derudder, 2016). Thus the enigma in the title of this section is premised on the 'urban' in urban demand. We have drawn profusely from Jacobs for our discussion and understanding of urban demand, but we also have noted in Chapter 2 that even our root source is problematic in regard to this concept: we found that Jacobs' analytic consideration of supply (innovation, import replacement) contrasts with import shifting (as urban demand) treated largely as a given.

The mysterious lack of cognition when it comes to urban demand is most clearly illustrated in the large array of UN meetings and conferences on climate change. These are classic cases of 'hidden in plain sight' – congregations of politicians, lobbyists, scientists, journalists and others locate in places that are hot stops of über urban demand. They choose to meet in large cities that meet participants' conference, hotel and restaurant requirements and are accessible by air from across the world. Notwithstanding that serious work is undertaken, these are urban jamborees that contribute to, and promote, the basic consumption function of the chosen cities. Urban demand is so very evident but appears unseen even when participating in it! Thus to house one of these demanding events is considered something of an economic coup for a city's economic development department, showcasing the city, not unlike hosting major sporting or cultural events.

Hidden or not, action has to be based on knowledge of the process being targeted for change. In this section we consider contemporary knowledge of urban demand – who knows what and why – and two future scenarios, both of which might significantly reduce climate change, but in drastically different ways.

Where is knowledge of contemporary urban demand to be found?

It is commonplace to argue that people migrate from rural to urban places because the latter provide a far greater range of opportunities. With their complex divisions of labour – many different jobs – new migrants can begin to make a new living, a stepping-stone to a better life. But the latter does not just mean work. Cities also have complex worlds of consumption, a different set of opportunities, myriad enticing products made available by the division of labour. This particular supply and demand interaction is behind the idea that, by moving to the city, migrants become different people. In this argument it is consumption that defines traditional differences between 'urban' and 'rural', not location per se.

Beyond such informal notions of consumption in cities, action to change über urban demand needs to be based on more critical analytical understandings of what it means and how it operates. This is where social science enters the argument. We can define social science simply as formal knowledge that seeks to understand how humans interact with one another in myriad different situations, building institutions to organise and manage those interactions. Cities represent the apex of complexity in these social interactions. By 'formal' we mean knowledge created by employing agreed practices for validating content and generating a systematic body of growing knowledge. Research is the operative process that accumulates knowledge; its practice defines the research frontier of social science. As such it is important to understand that there are two distinctive social sciences: university social science and corporate social science (Taylor, 2016). As their names indicate they differ in terms of where research takes place, but more importantly, they diverge in purpose.

The expectation is that when discussing social science we mean university social science. Certainly universities are where institutions of that name, locally, nationally and internationally, are organised. And this organised social science is centred on a tripartite division of its knowledge into economics, sociology and political science. These are the prime social science disciplines deriving from late 19th-century reform movements and consolidated in US universities in the mid-20th century (Wallerstein et al, 1996). As previously argued, their reform origins resulted in them becoming very state-centric in their knowledge production focusing on national societies, national economies and national politics respectively. This has meant that cities have been relatively neglected, certainly in terms of their extant importance. Disciplines discipline, researchers are certified (through successful PhD research) within disciplines. As noted previously, Jane Jacobs was never thus disciplined and took advantage of being a rare, un-caged social scientist.

What does all this mean for understanding urban demand? Unlike economic research on national demand, there is little or no systematic interrogation of the urban demand complex that we are concerned with. There are cultural studies on personal consumption, there are regional studies of demand in different sectors, business studies of consumption providers and planning studies of shopping centres, but no overall integrated body of knowledge of urban demand in university social science. Relevant knowledge of urban demand is largely concentrated in corporate social science.

What has happened is that the basic raw material of social science – social information, data and communication – have become commoditised. The world of corporate globalisation is denoted as 'informational', a communication society, captured in big data, all of which are far too important to be left to university researchers. Such knowledge is vital in enabling corporate globalisation to exist and grow. The result is myriad information-intensive firms creating corporate social science as a body of knowledge created as work for clients in the corporate world. A new economic sector has blossomed to provide this knowledge: advanced producer services doing financial, professional and creative work for clients. Much of their research is very specific, geographically recognising urban regions, and most definitely they do not neglect consumption. It is here that most knowledge of urban demand resides. In fact, we can say that there is a global Advertising-Big Data-Social Media complex roughly equivalent to the US Industrial-Military complex of the 1950s. The latter's power was focused on American production and its international projection; today's complex is clearly about consumption and is much more powerful through its role in expanding urban demand across the world. This is where knowledge, very practical knowledge, of urban demand is to be found.

There are two obvious implications of this knowledge location. First, the knowledge is private property, it can be kept confidential for commercial reasons, or it can be used publicly to promote the company holding the knowledge. In terms of the latter, information-intensive companies typically produce numerous reports and sponsor conferences and workshops – where cities often feature prominently – to illustrate their expertise in order to keep old clients and attract new ones. In this way corporate social science presents itself not unlike university social science contributing to a knowledge frontier, albeit in a more restrictive and purposive way. And the latter is the second key point. The Advertising-Big Data-Social Media complex is the complete opposite to the traditional university as an 'ivory tower'. The complex is commercial: its intrinsic purpose is simply to keep consumption increasing globally. Members of the complex might well research sustainability and sponsor 'green' conferences, but encouraging über urban demand is their fundamental raison d'être.

It should now be clearer why this chapter began with 'I wouldn't start from here'! Will it all end in tears for humanity? We should not shy away from this possibility: just such a scenario is envisaged next. But it is immediately followed by a more hopeful prognosis, although it must be admitted that this second scenario will need much more work to achieve than the first. This is where the action is vital.

Warlords: the urbicide solution?

From a conventional world-systems perspective, the challenge of anthropogenic climate change is treated as an additional element within an inevitable historical transition as the modern world-system runs out of ways of continuing to reproduce itself. Thus for Wallerstein (2004) the present is a special period: *kairos*, a critical time of decision. He describes a political bifurcation pitting the 'Party of Davos', working for a reconfigured hierarchical outcome, against the 'Party of Porto Allegro', working for an egalitarian alternative (2004: 87). In this analysis there is no consideration of climate change as an existential threat. But by bringing this into the argument something very different is created: an acute *kairos* with the radical response as much remedial as simply oppositional.

So what would the ending of our historical modern world-system look like? Actually we have a lot of knowledge about civilisation breakdowns: the crumbling of the established order creates a political void, which is filled by multiple local warlords. Operating in a system of might is right; simple military prowess provides a basic logic of social organisation. Cities lose their prime economic function as they become prizes to be won, looted for their wealth. The classic example is the fall of the Roman Empire in Western Europe resulting in the region's de-urbanisation. More relevant for our discussion is the demise of China, the last great world-empire, from the 1840s through to the 1940s. Here we see two processes operating simultaneously to undermine the established political order: alongside regional warlords exploiting the disintegration there were Western powers integrating the old empire into the modern world-system. This produced new urbanisation on the coast (for example, Shanghai) while there was urban decline in the interior of the empire. More generally we should be clear that these are not good times: for the Chinese, this is the curse of living in 'interesting times', for Europeans, the designation is 'dark ages'.

We are certainly living in interesting times: how dark might they get? In the case of the modern world-system the established order is the interstate system. This is a very uneven political order across the world; where there are 'failed states' in the periphery we already have warlords plying their might. But these are not a threat to the overall system. Much more relevant has been the rise of populism in richer states signalling the demise of the mainstream party competition of Right versus Left consolidated in the mid-20th century. Egotistic 'big men' have been destroying these liberal-democratic party regimes from Berlusconi in Italy in the 1990s through to Trump in today's USA. We are not arguing that these populist leaders are themselves preludes to warlords but, when combined with other processes undermining the modern state, they are an ominous development.

Although corporate globalisation has been enabled by states through their adoption of neoliberal policies, the resulting transfer of power to corporations has been detrimental to the workings of the liberal-democratic state. In addition to neoliberal tax reductions, large corporations manage their tax affairs transnationally. To prevent potential fiscal state crises, services to poorer citizens are reduced. Let down by the state, these deprived voters are fodder for populism. And populist governments are narrowly nationalist and thus are proving to be unreliable as partners in international deals on climate change. But curiously, their politics of promoting culture over economics does provide a pointer towards more climate-friendly politics. Liberal democratic politics involves parties effectively promising more consumption for votes – it's hard to envisage a party winning an election promising to reduce economic growth (and therefore consumption). The cultural politics of populism provides a break from this electoral formula – weak economic growth can be blamed on foreigners – but this in no way suggests a transition in the system, merely a political diversion. If there were to be a transition to a future rise of warlords it would be as a result of the populists' ultimately economic weakening of states. And it would also include populists' strong anti-metropolitan tendencies, resulting in the portrayal of big cosmopolitan cities as pits of iniquity rather than centres of creativity.

A new world of multiple warlords would likely result in mass urbicide, the death-knell of creative cities worldwide. Of course this would definitely reduce, indeed even eliminate, urban demand, but it is in no way a solution to anthropogenic climate change. The simple reason for this is that by the time we got to a breakdown of modern society to provide the necessary political void, climate change would have likely become too far down the route to an environment unsuitable for human habitation. Thus the warlords scenario is the potential real 'end of history'. The latter phrase was famously proclaimed after the collapse of the USSR: the victory of liberal democracy over illiberal communism was interpreted as the final triumph of American-led modern democracy, and there was nothing to follow. But warlords with their urbicide would literally end history: history begins with written texts; these are the creations of scribes – the initial advanced urban service – working in early cities; and it could conclude with a future concurrent worldwide ending of cities and writing.

Demand for what? Back to Jacobs

We derive two lessons from the unpalatable warlords scenario. First, the encouragement we get from Jacobs for bottom-up political action should not mean that we neglect governance: order can be repressive, but it is also enabling. Second, curtailing urban demand as we know it cannot mean destroying urban creativity; the latter has to be harnessed for alternative purposes. Despite her anarchistic tendencies, Jacobs understood the former, and provides the necessary insights to suggest ways of achieving the latter.

Jacobs' view on governance is explicitly covered in her study of moral syndromes. We have sight of these in two separate sources that provide an understanding into the evolution of her ideas (Lawrence, 1989; Jacobs, 1992). The labelling of her two syndromes is particularly instructive. The original 'working title' was 'Raiders and Traders', which relates to her identification of two ways of making a living: local use (raiding) of natural resources and non-local trading of products. Not satisfied with the first label she extended it to 'Raiding and Ruling' and then 'Government', which she found too narrow, and finally plumped for 'Guardian' (Lawrence, 1989). Clearly the move from emphasis on taking as 'Raiders' is very different from the emphasis on protecting as 'Guardian' syndrome. In this sequence she is admitting the basic need for order so that creativity can flourish through the Commercial syndrome. Furthermore, Jacobs (1992) stresses that each syndrome is of equal importance. In other words, some sort of social order mechanism is required for governance practices to complement bottom-up creative processes.

Although Jacobs (1992) had less difficulty in naming the Commercial syndrome, she still considered the label to be a little restrictive because she included work not facilitated through markets. This is particularly interesting when considering her earlier classic study on cities as vehicles of economic development (Jacobs, 1970). Make no mistake, her economic development is a process of economic growth, but her non-market insight opens up a whole new way of harnessing urban creativity. Ultimately she is writing about people making a living in the city, and there is much more to the complexity of the city than market competition. She gives examples of the urban demand mechanism where new work derives from old work in a range of non-market situations (in a hospital, a library and a museum). She also has a role for Guardians to encourage useful innovations and prevent the success of criminal innovations for which there is also an urban demand. More generally we would expect non-profit and non-governmental organisations, clustered in cities like Washington, DC and Nairobi, to be innovative and diffuse their new work to meet social needs in other cities. Clearly urban demand is not just an accumulation of commodities.

Putting these two syndrome arguments together we can note that political innovations can be similarly part of the urban demand complex. In fact, we have previously argued that historically city networks preceded city-states, the 'state' being invented to satisfy a demand for order in large fast-growing Mesopotamian cities, whose increasing cosmopolitan natures generated unwelcome social tension and conflict (Yoffee, 2005; Taylor, 2012a, 2013). Thus cities generate new Guardian work.

And so we return to urban demand. We can now ask, what demand? If, as Glaeser (2011) argues, cities are humanity's greatest invention, we certainly should not be persuaded to ditch them in the challenge of anthropogenic climate change. Put simply, we have never been in need of their creativity as much as we are today. It is not urban demand that needs stopping but rather, its particular incarnation as terminal consumption. Somehow we have to harness urban innovation and diffusion for alternative ends to current mass consumption. This requires not just a reinvention of 'the city'; it is a root and branch transformation of society through cities. Who, where and how can this be done?

The specific reality that is Chinese urbanisation

The progress of civilisation viewed through modern eyes had always been westward: starting in the ancient Middle East (Mesopotamia and Egypt), through classical Greece and Rome, to Renaissance (Italy) and Enlightenment (France) Europe, Britain's Industrial Revolution, and then across the Atlantic to the democratic USA as first global superpower and consumer champion. But in the second half of the 20th century there were early signs of a counter Eastern rising: first, Japan's economic success, followed by the so-called four little dragons – South Korea, Taiwan, Hong Kong and Singapore – and finally, the stirring of that great 'sleeping giant', China itself. In a complete reorientation, it became commonplace to label the coming new century variously the Pacific, Asian and even Chinese century (Frank, 1998).

Now well into this century, this broad categorisation is less used; attention is directed more specifically at the phenomenal rise of China. This is partly due to the contrasting fortunes of Japan and China over the last few decades, which have been acted out like some great Jacobs-inspired experiment. In one country, Japan, the dominant party of government used the wealth generated by the country's cities to bolster its rural electoral support in a policy of transferring investment away from cities. Osaka, Japan's second city, has been a notable victim of this process. The other country, China, experienced massive urbanisation accompanying its enormous economic development; here policy allowed concentration of investment in cities. The results are in. Japan, the initial harbinger of the Pacific century concept, has been in a state of continuing stagnation, even being out-grown by the old 'West' (USA and European Union), whereas China goes from strength to strength. While Japan's democratically elected political elite destroyed their previous momentum of post-war economic development, China's one-party political elite has presided over an economic miracle that really deserves to be called a miracle. This is exactly the opposite of what Western (aka liberal, modern) political commentators would have predicted (for example, Hutton, 2007), but it is the specific reality that is China today.

This interpretation of China's economic success is not the usual way the rise of China has been charted. The prime narrative is geopolitical, about great power rivalry – China versus the USA – and this has been recounted across the political spectrum (see, for example, Hutton, 2007; Li, 2008; Jacques, 2009). Such literature has little or nothing to say about China's massive urbanisation; for instance, in one case the only reference to Chinese urbanisation is historical, in the Song dynasty to be precise (Jacques, 2009: 76). It is certainly appropriate to mention China's urban legacy – it is in the East, not the West, where most great cities are to be found before the modern era (Taylor, 2013; Robinson et al, 2016) – but it is entirely inappropriate to ignore contemporary Chinese rural–urban migration, by far the largest movement of people in all of history (Miller, 2012)! In fact the total rural–urban migration in China in just one year, 1998, was larger than the whole of European migration to the USA between 1820 and 1920 (Campanella, 2008: 20). It is the latter movement that effectively created 'the West' that dominated the 20th century, but these trans-Atlantic flows of European people are truly miniscule compared with what has been happening in China. The contemporary changes in China have elicited many such superlatives, and one further example is equally enlightening. The amount of cement used in China in just three years (2012–14) is more than was used in the USA through all the 20th century (Shepard, 2015: 6). These extraordinary statistics are from an alternative narrative featuring cities, the one we employ. We contend that in considering anthropogenic climate change it is this second narrative that is essential, not one about disputes between US and Chinese governments, even when they are arguing at UN Climate Change Conferences.

The discussion on China proceeds in four parts. First, and following the previous superlatives, we consider China's urban boom in relation to similar past booms. The discussion then focuses on urban demand, and how that is operative in China's urbanisation, describing three distinct urban demand processes. The final part, recent Chinese policy, is specifically worrisome.

Size and speed really do matter

Since the Deng reforms of 1978, the resulting enormous changes in China have been manifest in economic development being definitively associated with urban growth. This should not be treated as a chicken-and-egg question; Jacobs (1970) tells us that economic development and urban growth are the same process. From this perspective, the Chinese experience can be compared to other examples of dynamic urban-economic change.

In the modern world-system there have been three classic examples of hugely successful economic developments based on rapid urban growth (Taylor et al, 2010; Taylor, 2013). These are Dutch cities in the 16th and 17th centuries, British cities in the 18th and 19th centuries, and US cities in the 19th and 20th centuries. These coincide with the three hegemonic cycles of the modern world-system centred on the Netherlands, Britain and the USA as a sequence of world hegemons (Wallerstein, 1984, 2004). At its peak ('high hegemony'), each country was not just economically dominant; they also defined the nature of being modern (Taylor, 1996a): the Dutch created mercantile modernity epitomised by Amsterdam (Barbour, 1963), the British created industrial modernity epitomised by Manchester (Briggs, 1963), and the USA created consumer modernity epitomised by Los Angeles (Scott and Soja, 1996).

Does China's huge economic surge mean that it is moving towards becoming the fourth hegemon of the modern world-system? Answering 'yes' would provide a neat argument in terms of simple continuity – each hegemon being larger than its predecessor – but it is not a credible answer. It is not just that the China's dynamic change is different – the other three cases are also very distinct from each other – there is also the crucial matter of the size and speed of China's economic development. First, the three recognised cases of hegemony each involved small proportions of humanity: minute in the case of the Netherlands, still tiny for the UK, larger for the USA, but still only a small fraction of the world's population. In contrast, China is pushing forward something like a sixth of humanity, and it is a proportion of a much larger total number of humans. Second, the rate of change is similarly on a completely different scale, from decades adding up to centuries for hegemonic cycles, to just a few years adding up to decades in the case of China. And these are much more than differences expected with a new hegemon; they bring up wholly new vital questions of sustainability and therefore the reproduction of the modern world-system. Past hegemons led our world-system to new modern worlds; it is not possible for China to have such a role because the system itself is in demise. Since China is coming to the fore in what Wallerstein (1984) refers to as the *kairos* phase (transition), this means its rise to importance is within an end time of critical decisions about future developments. As the modern world comes to be understood as just another historical interlude rather than the final climax of the human story, China reappears, not in its traditional political role as 'middle kingdom', but as, perhaps, humanity's main chance to reinvent planetary urbanisation.

If such a scenario is only partially correct, it does behove us to think through the way the urban demand complex has unfolded in China's unprecedented economic development and urban growth. Whither China's urban demand?

Urban demand I: a simple Jacobs process globalised

For Jacobs, economic development is defined as the creation of new work. And whatever else China represents, it has been the great generator of new work throughout its cities. This has been the result of the common mix of innovation and customised imitation with, as normal, the latter dominant. But it has not been based on the usual import replacement because the prime urban demand being satisfied is elsewhere – cities in the West. Thus the economic development has been Jacobs' (1970: 252–4) initial 'simple export-generating process'. In contemporary globalisation this has meant China producing a new global geography of supply and demand (Taylor, 2011).

Thus the internal development of cities in China is distinctively dependent on China's external commercial relations. A global space of flows has been created of raw material commodities in, consumer commodities out. At its simplest this has taken the form of a great worldwide 'triangle': the Global South provides the raw materials and the Global North buys the consumer products, with Chinese cities doing the work in the middle converting the production commodities into the consumer commodities for realising capital. In the Global South, Chinese cities have created their own world of economic dependencies; in Jacobs' (1984) terms these are simple supply economies or 'economic grotesques'. Campanella (2008: 295) describes the overall outcome of this process as 'From South America to Central Asia, China is literally consuming the world'. In the Global North, urban demand in the rich world is being satisfied by enormous imports from China. Thus we can add a second outcome, from North America to Western Europe, and for richer people worldwide, they are also consuming the world, albeit in much more pleasant circumstances.

Self-evidently, this unusual geographical location as an axle between external rural supply and external urban demand has been tremendously successful in terms of Chinese economic development. However, there is a very important, indeed frightening, point to make here: in Jacobs' urban development model it is NOT the initial export-generating process that produces the explosive urban growth that leads to economic development. And yet, without this awesome power of import replacement and shifting, Chinese cities have succeeded in rapidly generating immense quantities of new work in large numbers of cities. This must have something to do with the particular nature of rampant Chinese urbanisation. But how come? There has to be an additional source of urban demand in China's urban supply–demand complex.

Urban demand II: contra Weber

This second urban demand process is ensconced in Chinese cities: as well as the demand for raw materials for factories there has been huge urban demand for labour and materials to create the actual urbanisation, the new physical fabric – buildings and other infrastructure, for instance, in accommodating the unprecedented rural–urban migration through the colossal use of concrete mentioned previously. But how has this crucial, possibly one-off, massive urban demand come about and been satisfied?

Contemporary Chinese cities provide an interesting counter to classic Eurocentric thinking. Max Weber (1921/1958) has been the starting point for much writing on cities. He divided cities into two types – occidental and oriental, distinguished by their relative degrees of political freedom. Basically he argued that the latter had far less freedom than cities in Europe. Weber is thinking of the imperial free cities in the medieval Holy Roman Empire. He argued that this contributed to Europe's economic success in contrast to Asia's stagnation at the time of his writing. But the key point is that economic development requires commercial autonomy, not political freedom (Taylor, 2013: 236). In the hegemonic sequence previously discussed, the initial booming cities were found in central Holland, northern Britain and the US manufacturing belt, all geographically separated from their contemporary centres of political power in The Hague, London and Washington, DC respectively. Commercial urban autonomy comes in many different forms and in the contemporary Chinese case it is truly exceptional.

Lin et al (2015) show how urbanisation in China has become largely the result of the development policies of city governments. It is based on the long tradition of ownership of all land belonging to the government, initially imperial, now communist. Since China-wide urbanisation cannot be managed centrally, land development, including control of ownership of land, is devolved to local government (Zhang, 2003). This has provided city administrations with both the opportunity and the resources to promote large-scale urbanisation. Thus whereas in Western countries neoliberalism has resulted in a hollowing out of the state, especially at the local level, the commodification of land in China has been a crucial part of decentralising power to cities. And the resources made available to city governments appear almost inexhaustible: as well as benefiting from land leases for all urban developments within a city's administrative area, cities are also able to annex rural areas and accrue the consequent rises in value. In addition, development land, both built on and empty, is used as collateral for huge loans from local and state banks (McMahon, 2018). These are where the huge quantities of capital have come from to create the physical fabric of China's massive urbanisation.

This enormous locally constructed urban demand has taken some curious turns. China has become famous for its huge 'ghost cities', with houses and other infrastructure for hundreds of thousands of people, but which are currently left empty (Shepard, 2015; McMahon, 2018). Unlike traditional mining 'ghost towns' that disappear after the mining boom resides, China's ghost cities are pre-boom, speculations that future development will envelope them. This indicates a fundamental urban faith in the future found nowhere else in the world. Thus there is a unique 'future-urban demand' growth element built into the already outsized urban demand that is found today throughout China.

This is all contra Weber: contemporary Chinese cities have a commercial autonomy with concomitant resources that appear to be unprecedented in the long annals of urbanisations. Premised on the central government's need for social stability, other levels of government are given license to provide the necessary economic growth via new jobs, new infrastructures, new mass housing. Thus, contrary to Western perceptions, McMahon (2018: xix–xx) sets the scene for his analysis of China's economy thus: 'After years spent reporting on China's economy, I've learned that important change emanates from the bottom up, not the top down.' It is certainly not all good, injustices and corruption remain a problem: it is often more bulldozer Moses than community Jacobs. Nevertheless it has generated an amazing urban-economic outcome. It is this growth process that has offset the lack of import replacement and shifting, so central to Jacobs' model, but generally lacking in Chinese cities. It means that the export-led development, which Jacobs associates with slower urban growth, has been vitally boosted by massive local land development to create an alternative means of generating what is most definitely a case of 'explosive city growth'.

Urban demand III: a quick route to terminal consumption

Anyone doubting the power of the local commercial autonomy of Chinese cities in the reform era should consider this insight into the process provided by Fenby (2012: 171). He reports the Chinese leader and architect of the 1978 reforms, Deng Xiaoping, as saying that the urban transformation of the country 'was not anything I or any comrades had foreseen; it just came out of nowhere'. Obviously a remarkable admission, it shows that in national policy terms, China's urban revolution was hidden from plain sight, apparently beyond the ken of national elites. But this began to change with the 2008 global economic recession. Although they had operated successfully for several decades, China's central government now recognised the frailty of both the two urban demands reported previously that had stoked the country's economic development. The recession showed that the external urban demand could not be automatically relied on. And the internal urban demand, reliance on city land development for urban growth, is not sustainable over the long run, as McMahon (2018: 59) has clearly shown.

So in 2012 urban development finally makes its way into the national five-year plan. One policy initiative is obvious: to encourage the domestic economy to generate an internal urban demand for Chinese-made commodities previously exported (that is, production from Chinese cities to be consumed in Chinese cities). This makes a lot of political sense for China's communist rulers, both by garnering a growing middle-class support for the Communist Party, and by triggering Jacobs' import replacement and shifting mechanism to continue economic development as new work. But it does not make any sense whatsoever in terms of anthropogenic climate change. In fact, quite the opposite – such an urban economic turn in China is globally calamitous.

The logic behind this extreme forewarning needs to be spelt out precisely:

- Jacobs' (1970: 260) economic development model is a spiral of change based on 'two reciprocating systems of urban growth'.
- The first process is a slow growth mechanism based on a simple export multiplier.
- This feeds into a second process, a rapid growth mechanism that is import replacing and shifting.
- It is the latter process, so far not developed in Chinese cities, that has historically generated Jacobs' famous periods of 'explosive city growth'.
- Thus what we have in China is the first mechanism converted to a means of generating fast growth as explained previously, and now this

is feeding into a new economic phase that history tells us massively enhances the urban growth that went before.

In other words, following on from the largest and most rapid urbanisation in history, there are to be added multiple bursts of new, enormously enhanced, urban growth in cities throughout China. The consequent amassing of additional urban demand is simply mind-boggling. This is the moment that forces us into a change in terminology. Now it is confirmed unequivocally, from a global perspective – what we are facing is nothing less than terminal consumption.

What next? Making fundamental urban change

This chapter is labelled 'Action' and yet, actually, we have provided very little indication of the vital action that is needed to douse terminal consumption. However, throughout the book we have faced up to the daunting challenge of the climate emergency to provide a radical framing of the need to respond through cities. In this we have used Jane Jacobs as our guide whose key mechanism of economic development generates ever-expanding urban demand, culminating in contemporary China's unprecedented urban boom. It follows that the societal transition required to deal with climate change must take the form of fundamental urban change. This means turning urban demand inside out. Easy to say, but what to do? We conclude this book by presenting four arguments as aids to answering this question: What politics? What policy? What city? What geography?

What politics? Superseding capitalism on the cheap

We have discussed politics previously, and advocated a bottom-up approach. Hence we don't think that delivering fundamental urban change will be simple top-down policies by city mayors, charismatic or not. However, in this discussion we change tack from a focus on political practice to the content of a new politics. Generating a politics to combat the climate emergency will not take place in a radical politics void. Our trans-modern thinking will have to operate within radical politics long generated through the modern world-system. Traditionally, radical modern politics operated within different national frameworks of class conflict in which various parties of the Left challenged capital (or at least its socially pernicious effects) and were opposed by parties of the Right supporting capital. The result has been multiple national politics moulded largely by production and supply, with consumption and demand generally emerging as just policy outcome.

This radical focus on division of labour – economic classes in production – is at the heart of Marxist political theory. However, as previously noted, there are signs of change here. David Harvey, arguably the most important Marxist political economist in Anglo-American social science, is reinterpreting *Das Kapital* for the 21st century. He uses the water cycle, where things change from gas through liquids to solids, as an analogue of the cycle of accumulation from production through circulation to consumption. Both the water cycle and the cycle of accumulation show metamorphoses, in which things change in shape and form. What we want to draw from Harvey, forgetting much of the richness of his analysis, is one of his conclusions that maybe radical scholars have emphasised too much, the importance of *Kapital Volume One*, on production, at the expense of circulation and realisation, the subject of *Kapital*'s two other volumes. This over-emphasis has had direct consequences for political struggle, of framing Left versus Right, as capitalist against worker, realised more generally as the importance of trade unions to Left parties. Harvey more than hints this might be an over-emphasis on production. This is, of course, a move towards our position; consumption is needed for realisation of capital (profits) and is facilitated by circulation.

However, the main revision of traditional radical politics has come from the addition of other conflicts, particularly in gender, race, ethnicity, sexuality and the environment. Whether interpreted as broadening or fragmenting, this is the context in which any new radical focus on tackling urban demand begins. We use the work of Jason Moore (2015) to ground this radical variety. He has placed geographical frontiers at the centre of studies investigating the historical ecological impact of capitalism. In the previous chapter we disputed his revision of the concept of the Anthropocene to the Capitalocene. Our difference with him on nomenclature derives from his focus on supply (frontiers) rather than demand, whereof we have suggested the Urbanocene. But his analysis of supply frontiers complements our urban demand argument, and we interweave them to search out a new politics for climate change. For instance, in our interpretation, the stagnant urban world of communist China in the third quarter of the 20th century turned out to be the last great frontier of capitalist expansion after 1978. This case and Moore's historical frontiers are where society and the rest of Nature are intensely entangled; they are forms of violence, taking things without paying their real value. We can equate them with Jacobs' (1984) city-created supply regions, but Moore adds the all-important world ecology dimension.

We use Moore's co-authored treatise *A History of the World in Seven Cheap Things* (Patel and Moore, 2018). Patel and Moore identify seven forms of exploitation and appropriation – the 'cheap things' of their title – that have characterised the expansion of capitalism. These are cheap nature, cheap money, cheap work, cheap care, cheap food, cheap energy and cheap lives. All these 'things' are crucial elements of changing urban demand complexes. They are cheap because although being essential to capital accumulation, these goods, people and environment remain absurdly under-valued in capital expenditure and are, therefore, a key source of accumulation. Nevertheless, each case of cheapness has generated its own opposition. We deal briefly with each 'cheap' in turn, defining, illustrating and linking to urban demand and radical politics.

Cheap nature. Three things underpin the violence of cheap nature. The first is the triumph of the Enlightenment Project, broadly the rise of science and technology to control Nature in order to exploit it. This requires a separation of people and environment while simultaneously having people as part of environment. The contradiction was expressed by Smith and O'Keefe (1980) as the 'production of nature', which saw nature not just as resources to consume, but also as a sink for pollution. The carbon crisis is this sink writ large as a global dustbin for combustion and where the real costs have only recently begun to be appreciated. Urban areas are the most conspicuous in their pollution, not least contemporary Chinese cities. The environmental movement, often grown from a rural base, has evolved to generally oppose cheap nature, albeit frequently from a population crisis position. Remove rural focus and Malthusianism, and the ensuing environmental politics is central to a politics of urban demand.

Cheap money. This is not money for exchange but money to accumulate. Money acts at a distance, smashing through new frontiers. As told by Wallerstein (1974), the modern world-economy began as a trans-Atlantic economy initially boosted by Spanish bullion from the Americas. Epitomised by the 'silver mountain', Potosi, the bullion defined the early capitalist world-economy as economically inferior to the leading world-empire in China, which is where much of the silver finished up (Frank, 1998). Subsequent European imperialisms turned this relation around by the 19th century. Cecil Rhodes, standing in Cape Town, South Africa, looking north to Cairo, Egypt, summed it all by saying the frontier expansion was for commerce, Queen and God, in that order: World Money (now gold) for World Nature (now tropical ecosystems).

Today, cheap money has become ubiquitous. In the East, manipulating the currency market to keep the currency cheap has been central to China's recent economic growth. In the West since the 2008 banking crisis 'quantitative easing' (printing money) has saved the banks. The opposition to this is many and varied, most explicitly led by 'Occupy Wall Street' movement.

Cheap work. This needs little amplification for it is the essence of trade union activity referred to previously. Beyond traditional wage negotiations and conditions, in this era of austerity, it includes struggles over the minimum or living wage, zero-hour contracts plus the maintenance of employment rights, including pensions. It is also at the centre of important political debates on privatisation (moving workers from public to private employers), rising inequalities (the demise of trade unions and corporate pay policies), immigration (refugees and economic migrants) and the future of employment itself (AI, etc). Accompanying these growing challenges, urban economic structures have moved away from traditional manufacture towards services, a key reason for reduction of trade union membership. Add to all this capital's global search for cheap labour over the last half century: redundancies in deindustrialising richer countries replaced by very cheap labour in poorer countries, culminating in China dominating world manufacturing. Hence this longest standing Left politics is being diluted.

Cheap care. This means costs of social reproduction, looking after children, looking after family, looking after the old folk – work that is largely done by women, work that is largely unpaid or, at best, underpaid. For Patel and Moore (2018: 123), 'Writing about a history of work without care work would be like writing an ecology of fish without mentioning the water.' This example of cheapness takes two main forms. In many of the poorer parts of the world, households are geographically split, with paid labour of emigrants (usually men) far away from the rest of the household in the home farm, where mostly women, children (some future emigrant workers) and the elderly (some former emigrant workers) live. Wallerstein (1984) calls these 'semi-proletariat households' because reproduction costs (on the farm) are not included in the paid wage (in the city). In richer parts of the world similar use of the household to provide care (but without the geographical separation) has been a major stimulus of the feminist movement, with ongoing political issues over care not being paid for by capital. Waves of feminist theory, discourse and intervention continue to be key sources of what we have called 'unthinking'.

Cheap food. As discussed in previous chapters, Jacobs (1970) has food production as one of the first urban innovations, and later uses it to name the 'Plantation Age' as indicating the construction of uniform landscapes at the altar of urban demand (Jacobs, 2004b/2016). She does not recognise the intensification of this process with the capitalist world-economy, but as agricultural production moved from political control to being market-led, the ecological destruction has moved from local and regional to global in scale. From the 'second feudalism' in Eastern Europe in the 16th century to provide grain for booming Dutch cities, through North Atlantic fishing to feed Britain's industrial towns and cities, and Chicago coordinating the 'Corn Belt' to produce pig meat for North East US cities in the 19th century, to today's corporate ranching, industrial fishing and factory farming, the food industry has operated at the expense of ecologies worldwide to produce cheap food. This has facilitated lower wages and therefore contributes to capital's profits. The real costs are massive: moving first carbohydrates and then meats to cities has become energy and chemically intensive, to say nothing of the profligate uses of water. This activity today probably accounts for over one-third of greenhouse gases. Unsurprisingly, the food industry has generated a range of opposition, originally from small family farms to today's organic soil movement through to the growing influence of veganism.

Cheap energy. All three authors grew up surrounded by family who worked in the northern coalfields of England. After wood as the basic source of pre-capitalist energy had run out, land rights, including mineral rights, of traditional landowners enabled them to profit from coal and later oil and gas. Little remained in the hands of the state; most was private accumulation. It was a private accumulation largely based on the assumption that hydrocarbons were nature's bounty, a free gift, the only cost to the landowner being gathering and moving to market. Coal was initially directly linked to steam, then gas and electricity, but the fuel switch did not forsake the axiom of cheap energy. When hydrocarbons seemed threatened, either by monopolies other than the Western world (for example, OPEC) or by the threat of resource extinction, new cheap energies were promoted: nuclear power arrived with a claim that it would be too cheap to metre! Of course, the project to keep the price of fuels relatively cheap comes at a huge un-costed cost. Two hundred years' worth of emissions have been a prime cause of increasing greenhouse gases that have accelerated climate change. While there is some opposition, from environmental movements to NIMBY actions, there is no unified push against cheap energy.

Cheap lives. We need to be specific here because we have already dealt with women in households, so what is the focus? Black lives. Some 12.5 million Africans were shipped across the Atlantic in the slave trade, of whom 10.7 million survived. The 4 million slaves alive in the USA just before the Civil War were, according to Srinivasan (2017), worth some US$2.8 billion. He contrasts this with the total value of the railways, the single most important capital asset, at US$1 billion and the Federal Government's total annual turnover of US$69 million in 1859. People as property: no wonder the Confederacy went to war to preserve this 'human capital'. This is reinforced by the UK example, freeing slaves in its early 19th-century empire. In 2018, the UK Treasury tweeted that it had finally paid off a debt for compensation from the end of slavery in 1831. It means that, for all the authors' working lives, part of our taxes had gone to recompense slave owners' families in Britain. In complete contrast, freedom for slaves simply reinvented the racial structure as a class structure, although continued denial of rights saw the struggles as ethnic rather than class. With reference to the Nature/society tension, this emerged as the struggle for environmental justice, a rejection that blacks must live in the most polluted and polluting environments. But black struggle both in the USA and the anti-colonial movement was probably the most successful of struggles against capitalism in the late 19th and 20th centuries, whose lessons must be captured.

There are two related threats common to all seven cheap things. First, there are limits; ultimately cheap things are not inexhaustible. This is how Wallerstein (2004) sees the capitalist world-economy building towards its demise. Second, and feeding off the first, as things get tighter, they become more violent and thereby stimulate more political opposition. This is the opportunity for a new trans-modern politics harnessing these disparate modern radical movements and oppositions together to face the common threat of terminal consumption. Thus the movements and campaigns identified previously are the critical context, the political raw material, for mobilising a new radical response to the climate emergency. As always, this will be created in a variety of ways by different groups across the world, but the communality of the threat has to be recognised – we are all in this together, even if some are initially hurting more than others. This will be a particular challenge to the individuation of the strong movements centred on gender and race or ethnicity. But a 'rainbow coalition' of radical movements and critical unthinking, both in opposition to capital and ignoring its modern enticements, can be the only basis of turning mass urban demand inside out.

What policy? An adaptation spectrum

Politics begets policies, practical ways of implementing ideas and attaining given goals. In contemporary policy-making for tackling climate change there are frameworks specifically developed to this end at community and regional scales (O'Brien and O'Keefe, 2014). Here we address these practical concepts as tools for a radical politics operating at a global scale.

Confronting risks from climate change are commonly divided into two distinct strategies, namely, mitigation and adaptation. Mitigation essentially is to stop physical nature battering human beings. It usually involves a high level of capital and technology, such as dams and coastal defences. Thus it tends to be capital-intensive, requiring governments to commit large amounts of money upfront. At the global scale this is called geo-engineering, wherein scientists explain how scientists can save the world. If it sounds like science fiction, that is because that's what it is. Eschewing the complexity of the Earth's physical systems, mega-projects like reflecting back sunlight or changing the oceans to absorb more carbon, assume climate change is a simple problem. The unintentional collateral damages don't bear thinking about. Fortunately most scientists agree; mitigation has limited traction in policies on climate change beyond specific local needs. Such a strategy is necessary for creating resilient cities.

Adaptation is different; it is where human beings adjust their actions to counteract the threat of a changed nature. Unlike mitigation, adaptation usually requires not a one-off payment but a revenue stream that can fund changes in livelihood and lifestyle. In general, governments do not like calls on revenue accounts because revenue accounts place pressure on the tax base. Globally adaptation also provides a different challenge since many adaptation measures will have to be taken in poorer developing countries that, by definition, have a weak tax base. The global policy challenge is therefore difficult where richer countries who have had, and still do have, higher carbon emissions, will be asked to bear the burden of finance to help poorer countries. The snarl up in current climate negotiations is largely about this issue. But do we want to adapt to global climate change in this sense? What are we actually sustaining? Obviously at this scale we do not want to employ policies that sustain the human and physical elements of anthropogenic climate change that have got us where we are today. However, adaptation strategies are common at the urban scale in order to create sustainable cities. Again, this is a necessary policy for local city residents.

Instead of contrasting mitigation and adaptation, the former is sometimes regarded as a subset of adaptation. Following along this line of thought we can consider adaptation as a spectrum of policy options ranging from basic mitigation to some sort of responsible transformation. In this argument we find the 'resilient city' (mitigation) at one end, the 'sustainable city' (traditional adaptation) at the centre, and what we will call the 'posterity city' (transition for future generations) at the other end (see Figure 5.1). The latter involves being much more proactive in regard to our relations with the rest of nature. This is stewardship, work directed at preserving and nurturing the environment as a human habitat for generations to come. This is a radical adaptation policy raised up to the scale of cities in general as a planetary urbanisation; it points towards developing deep-seated social change, setting forth a necessary new path for humanity.

Figure 5.1: The adaptation spectrum

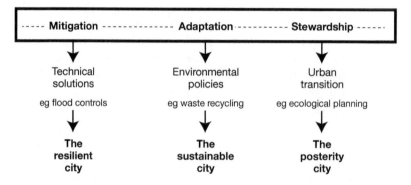

What city? Reinventing the city as stewardship

We are sure of one thing: the city has to be reinvented. Of course, dynamic cities are incessantly reinventing themselves as they add new work to the city economy. But there are rare occasions when fresh bundling of new work has repercussions far beyond a specific city economy: its success shows the way for other cities to create novel economic complexes. For instance, such a new-fangled urbanism was pioneered by Los Angeles – the extensive, car-dependent consumption city – harbinger of mass suburbanisation of the second half of the 20th century. In previous centuries Amsterdam pioneered a new mercantile city and Manchester a new industrial city. Campanella (2008) thinks the Chinese have reinvented the city in the 21st century; certainly the urban process is very different in China, as we have briefly shown. But what is needed for a fundamental reinvention is much more than a speeding up of an existing urban process reflecting a particular economic, cultural and political set of circumstances.

For a basic reinvention, the urban process keeps what is good about the city (dynamism) while eliminating its destructive potentials (lethargy). Hence an economic no-growth strategy to protect the environment is not what we are looking for because it would remove the creative side of the city that derives from vibrant agglomeration and cosmopolitanism. As Jacobs tells it, stagnant cities are the greatest polluters of all. A dull, austere, disconnected urban world is neither attractive nor functional. In fact, China experimented with a form of this anti-urban approach in the 1960s Cultural Revolution: an economic disaster, it made China the only country in the world to de-urbanise in a decade that otherwise had worldwide rapid urban growth.

Humanity's creativity has blossomed through its cities, but current consumption-led growth is simply delinquent. Turning this model inside out we find the posterity city from our adaptation spectrum. This radical transition would not be a collection of myriad environmentally friendly policies and actions in an urban context – what Jacobs (2000) referred to earlier as a 'thing' theory of development – rather, there has to be a new process of city-making. The whole raison d'être of the city has to be stewardship of Nature in line with humanity's prime purpose of preventing terminal consumption. Thus there will be new work to make market-enduring stuff and services that develop and maintain a vibrant nature including humanity: a dynamic division of labour ensures continuity of city complexity. And this will involve an urban demand being functionally satisfied, but one that realises a very different outcome to the present. Posterity cities can be said to exist when a first round of stewardship new work is transmuted into routine old work having been superseded by the latest stewardship new work: a reinvention of the city is thereby confirmed.

Posterity cities focus on city life, not city space. The latter invokes mastery of Nature as typically seen in the work of architects and planners: this so-called mastery is an illusion that has helped to get us to where we are today. Thus deep stewardship should not be confused with custodianship – humans 'looking after' Nature – because policy must be holistic, no longer separating humans from Nature. Prioritising how we live is about the husbandry of all Nature, ecologic work that inevitably varies over time and space. In a historical context, a 'steward' is not 'the boss'; rather, he or she is a specialist practitioner who oversees resources for future generations. This is what cities will be required to do. Different ways of urban life – there cannot be a one stewardship process fits all – generate a geography of stewardship ways and means. Worldwide differences will follow diverse urban demands reflecting varieties of environmental needs and cultural contexts.

What geography? Chinese cities' role in planetary stewardship

Previous reinventions of the city have had a specific geographical source. In our current globalised times this may not be so spatially clear-cut because of the geographical variety just alluded to: there are and will be myriad ways of transforming cities across the world. But as we have indicated in this chapter, a prime locale for reinvention will have to be China. This is not just because of the scale and speed argument mentioned earlier, although these are very important. And it is certainly not a matter of geopolitical transition: the West has had its chance, its messed things up, somewhere else has to take over the leadership, hence China. Rather, Western cities obviously remain important in the stewardship of the Earth, but these are the places where modernity, culminating in mass urban demand as terminal consumption, was invented and developed, and therefore it is where the toxic urban processes are so deeply embedded. More overtly locked in, this likely makes Western cities that much harder to turn around.

However, following Jacobs, we are predisposed to strong bottom-up action and at first this seems somewhat problematic in China, ruled by its powerful Communist Party (CCP). Although Jacobs concedes the need for good governance, one that is tolerant of contrarian ideas to allow unplanned creativity, this does not sound like a job for a one-party state. In fact, current urban development in China has been severely criticised precisely for its relative lack of creativity: Shepard (2015: 18) refers disparagingly to the endless production of 'generic cities'. This sounds very much like Jacobs' sterile plantation outcomes, where vitality is squeezed out of development. Thus focusing on China appears to have brought us back to where we began this chapter – we wouldn't start from here!

Not so. As we have previously indicated through McMahon's (2018) long-term observation of contemporary China, the rule of the CCP is by no means the sort of top-down governance as commonly understood. Lim (2019: 5) adds detail to this argument: the party has a long tradition of decentralising authority from military necessity before becoming the party of Chinese government in 1949. This tradition has continued in different forms down to the present. Lim focuses on the policy of new strategic areas that has been developed after the shock of the 2008 global crisis. Realising inherent problems stemming from three decades of rapid economic growth, the CCP's response has not been to search for new foreign recipes to impose on the country; rather, it decided to test potential solutions for specific problems domestically through innovative experimentation (Lim, 2019: 3–6). More than a dozen such strategic areas have been selected. For instance, Chongqing focuses on finding solutions to the inequities rural migrants face by experimenting with rural–urban integration. The point is that these are open-ended experiments in a politics of multiple scales, with cities and regions lobbying and negotiating with central government for the resources to carry out nationally important local initiatives. This is not the typical political decentralisation we find in federalism – the CCP centre still formally controls who gets what – but it encompasses policy flexibility. In addition, city governments and local businesses are experts at getting around centrally imposed rules that hinder their needs (McMahon, 2018: 24). There is certainly the raw material here for a city-led urban transition.

On a more practical note, we can add that the CCP itself is certainly not immune from fundamental changes; in fact, it has a strong track record in this respect. For all political regimes, staying in power seems to be the common driver, and the CCP has been particularly successful in this pursuit. An obvious case is the 1978 reforms when the Chinese government moved to make their country economically more like their successful Asian neighbours. This meant physically developing their cities America-style, reflected today by superhighways cutting through neighbourhoods while being snarled up by traffic: not a good long-term choice. A change of governance direction to become less like the USA in the not too distant future cannot be ruled out: another 1978-level turnaround would be required. And, of course, promotion of a new un-American practice might well be particularly attractive. China's current government is very mindful of its difference from the West, but its great urban renaissance of recent decades should not be seen as a simple stepping-stone for replacing the USA as 'world number 1' in the way geopoliticians might argue. China's political mentality is more long term: this makes China specifically open to seeing modernity as just another historical interlude in a way that is much more difficult for the USA, and the West in general, as the creators of being modern.

These lines of thinking can lead to numerous future scenarios, one of which would mean China's cities having a leading role in combating anthropogenic climate change. The key characteristic of posterity cities is maintenance of what Jacobs (2000) terms 'dynamic stability', a viable process of creativity change. This can be translated into traditional Chinese thinking as 'dynamic harmony', the ultimate trans-modern city/state nexus: Jacobs meets Confucius! If ever there was a topic in need of innovative experimentation, this is it: one or more strategic new areas testing stewardship ideas. The problem looking for a solution is how current global urban policy for economic growth can be reset as urban planetary policy for creating a posterity city.

Here we can return to our thesis/antithesis mode of reasoning. Successful experimentation would result in Chinese cities embarking on a route to becoming the obverse of US consumption cities. In the early 20th century the latter developed new professional urban services – management consultancy and advertising – to grow mass consumption, by making stuff more affordable while concurrently generating new wants for said stuff. This has culminated in today's Advertising-Big Data-Social Media global complex that exists precisely to keep the mass consumption bonanza going. In contrast to these architects of terminal consumption, Chinese cities in the 21st century would need to create a real alternate: the development of new professional urban services enabling city stewardship. This might be a Stewardship-Big Data-Social Media complex to generate an alternative urban demand, reversing the mass clamber for ever more transient stuff.

Creating and maintaining a mass stewardship want is the major ask of posterity cities, Chinese or otherwise. No one world region will monopolise fundamental responses to the climate emergency: if Chinese cities do embark on a development path as suggested previously, cities in the rest of the world will not be standing still while extreme weather conditions multiply. Ultimately the core nature of posterity cities will likely be a result of synthesis of reforming US-style consumer cities and their radical Chinese obverse. A Jacobs' future might be a suite of vibrant posterity cities acting as 'mother cities' across the world, creating alternate developments and steering away from terminal consumption.

References

Abu-Lughod, J. (1989) *Before European Hegemony: The World System AD 1250–1350*, New York and Oxford: Oxford University Press.

Agnew, J. (1993) 'The territorial trap: The geographical assumptions of international relations theory', *Review of International Political Economy*, 1: 53–80.

Algaze, G. (2005a) *The Uruk World-System: The Dynamics of Expansion of Early Mesopotamian Civilization*, Chicago, IL: University of Chicago Press.

Algaze, G. (2005b) 'The Sumerian take-off', *Structure and Dynamics: eJournal of Anthropological and Related Sciences*, 1(1): Article 2. Available at: https://escholarship.org/uc/item/76r673km (accessed 22 October 2019).

Amin, A. and Thrift, N. (2002) *Cities: Reimaging the Urban*, London: Polity.

Amin, S. (1990) *Delinking*, London: Zed.

Angus, I. (2015) 'When did the Anthropocene begin ... And why does it matter?', *Monthly Review*, 67(4): 1–3.

Arrighi, G. (1994) *The Long Twentieth Century*, London: Verso.

Askins, K. and Pain, R. (2011) 'Contact zones: Participation, materiality and the messiness of interaction', *Environment & Planning D*, 29: 803–21.

Barbour, V. (1963) *Capitalism in Amsterdam in the Seventeenth Century*, Ann Arbor, MI: University of Michigan Press.

Batty, M. (2013) *The New Science of Cities*, Cambridge, MA: MIT Press.

Belich, J. (2009) *Replenishing the Earth: The Settler Revolution and the Rise of the Anglo-World, 1783–1939*, Oxford: Oxford University Press.

Belussi, F. and Caldari, K. (2009) 'At the origin of the industrial district: Alfred Marshall and the Cambridge School', *Cambridge Journal of Economics*, 33: 335–55.

Berman, M. (1988) *All that is Solid Melts into Air*, London: Penguin.

Bookchin, M. (1995) *From Urbanization to Cities*, London: Cassell.

Boserup, E. (1965) *The Conditions of Agricultural Growth*, New York: Aldine.

Brand, S. (2010) *Whole Earth Discipline*, London: Atlantic Books.

Braudel, F. (1972) 'History and the social sciences: The *longue durée*', in P. Burke (ed) *Economy and Society in Early Modern Europe*, London: Routledge Kegan Paul, 25–53.

Braudel, F. (1981) *The Structures of Everyday Life*, London: Collins.

Braudel, F. (1982) *The Wheels of Commerce*, London: Collins.
Braudel, F. (1984) *The Perspective of the World*, London: Collins.
Brenner, N. (ed) (2014) *Implosions/Explosions: Towards a Study of Planetary Urbanization*, Berlin: Jovis.
Briggs, A. (1963) *Victorian Cities*, London: Penguin.
Brooke, J.L. (2014) *Climate Change and the Course of Global History*, Cambridge: Cambridge University Press.
Brugmann, J. (2009) *Welcome to the Urban Revolution*, London: Bloomsbury Books.
Bulkeley, H. (2013) *Cities and Climate Change*, London: Routledge.
Byrne, D. (1998) *Complexity Theory and the Social Sciences*, London: Routledge.
Campanella, T.J. (2008) *The Concrete Dragon*, New York: Princeton Architectural Press.
Carter, H. (1983) *An Introduction to Urban Historical Geography*, London: Arnold.
Castells, M. (1996) *The Rise of the Network Society*, Oxford: Blackwell.
Castells, M. (1999) 'Grassrooting the spaces of flows', *Urban Geography*, 20: 294–302.
Castree, N. (2015) 'New thinking for a new Earth', *Entitle Blog*. Available at: http://entitleblog.org/2015/11/30/new-thinking-for-a-new-earth/ (accessed 1 December 2015).
Chandler, T. (1987) *Four Thousand Years of Urban Growth: An Historical Census*, Lewiston, NY: Edwin Mellen Press.
Chase-Dunn, C. and Manning, S. (2002) 'City systems and world-systems: Four millennia of city growth and decline', *Cross-Cultural Research*, 36: 379–98.
Childe, V.G. (1950) 'The urban revolution', *Town Planning Review*, 21: 3–17.
Choe, S-C. (1998) 'Urban corridors in Pacific Asia', in L. Fu-chen and Y.-M. Yeung (eds) *Globalization and the World of Large Cities*, Tokyo: United Nations University Press, 155–73.
Cichello, A. (1989) 'In defense of Jane Jacobs: An appreciative overview', in F. Lawrence (ed) *Ethics in the Making: The Jane Jacobs Conference*, Atlanta, GA: Scholars Press, 99–169.
Clement, C.R., Denevan, W.M., Heckenberger, M.J., Junqueira, A.B., et al (2015) 'The domestication of Amazonia before European conquest', *Proceedings of the Royal Society B*. Available at: http://rspb.royalsocietypublishing.org/ (accessed 29 September 2015).
Climate Transparency (2018) *G20 Brown to Green Report*. Available at www.climate-transparency.org/online-launch-of-brown-to-green-report-2018 (accessed 15 December 2018).

Cronon, W. (1991) *Nature's Metropolis: Chicago and the Great West*, New York: Norton.

Crosby, A.W. (2004) *Ecological Imperialism: The Biological Expansion of Europe, 900–1900*, Cambridge: Cambridge University Press.

Crutzen, P.J. (2002) 'Geology of mankind – The Anthropocene', *Nature*, 415: 23.

Cunliffe, B. (2008) *Europe between the Oceans, 9,000 BC–AD 1000*, New Haven, CT: Yale University Press.

Curtin, J. (2018) *The Paris Agreement Versus the Trump Effect: Countervailing Forces for Decarbonisation*, Dublin: Institute of International and European Affairs.

Curtin, P.D. (1984) *Cross-Cultural Trade in World History*, Cambridge: Cambridge University Press.

Dalby, S. (2018) 'The Anthropocene thesis', in M. Juergensmeyer, S. Sassen and M.B. Steger (eds) *The Oxford Handbook of Global Studies*, Oxford: Oxford University Press, Chapter 11.

Dauvergne, P. (2008) *The Shadows of Consumption*, Cambridge, MA: MIT Press.

Davidovich, V.G. (1974) 'On patterns and tendencies of urban settlement in the USSR', in G.J. Demko and R.J. Fuchs (eds) *Geographical Perspectives in the Soviet Union*, Columbus, OH: Ohio State University Press, 611–33.

Desrochers, P. and Hospers, G.-J. (2007) 'Cities and the economic development of cities: An essay of Jane Jacobs' contribution to economic theory', *Canadian Journal of Regional Science*, 30: 115–30.

Diamond, J. (1997) *Guns, Germs, and Steel: The Fates of Human Societies*, New York: W.W. Norton.

Duranton, G. (2017) 'The death and life of great American cities/The economy of cities', *Regional Studies*, 51: 1871–3.

Ehrlich, P.R. (1971) *The Population Bomb*, New York: Ballantine.

Ellis, E. (2016) 'Involve social scientists in defining the Anthropocene', *Nature*, 540: 192–3.

Fairbridge, R.W. and Agenbroad, L.D. (2018) 'Holocene epoch', *Geochronology*. Available at: www.britannica.com/science/Holocene-Epoch (accessed 22 October 2019).

Faludi, A. (2002) *European Spatial Planning*, Cambridge, MA: Lincoln Institute of Land Policy.

Fenby, J. (2012) *Tiger Head, Snake Tails*, London: Simon & Schuster.

Flint, A. (2009) *Wrestling with Moses: How Jane Jacobs Took on New York's Master Builder and Transformed the American City*, New York: Random House.

Foresight (2008) *Powering Our Lives: Sustainable Energy Management and the Built Environment*, London: Government Office for Science.

Frank, A.G. (1969) *Latin America: Underdevelopment or Revolution*, New York: Monthly Review Press.

Frank, A.G. (1998) *ReOrient: Global Economy in the Asian Age*, Berkeley, CA: University of California Press.

Froy, F. (2018) 'Is new work really built from old work? And if so, what does this mean for the spatial organisation of economic activities in cities?', *Proceedings of the Conference 'Jane Jacobs 100: Her Legacy and Relevance in the 21st century'*, Delft, Netherlands: TU Delft, 209–15. Available at: www.researchgate.net/publication/325763681 (accessed 16 July 2018).

Funtowicz, S. and Ravetz, J.R. (1991) 'A new scientific methodology for global environmental issues', in R. Constanza (ed) *Ecological Economics: The Science and Management of Sustainability*, New York: Columbia University Press, 137–52.

Gamble, C. (2007) *Origins and Revolutions: Human Identity in Earliest Prehistory*, Cambridge: Cambridge University Press.

Gibson-Graham, J.K. (2008) 'Diverse economies: Performative practices for "other worlds"', *Progress in Human Geography*, 35: 613–32.

Giddens, A. (2009) *The Politics of Climate Change*, Cambridge: Polity Press.

Gignoux, C.R., Henn, B.M. and Mountain, J.L. (2011) 'Rapid, global demographic expansions after the origins of agriculture', *PNAS*, 108: 6044–9.

Glaeser, E. (2011) *Triumph of the City*, London: Macmillan.

Graeber, D. (2011) *Debt: The First 5,000 Years*, New York: Melville House.

Gregg, S.A. (1988) *Foragers and Farmers: Population Interaction and Agricultural Expansion in Prehistoric Europe*, Chicago, IL: University of Chicago Press.

Goldsmith, S.A. and Lynne, E. (2010) 'Eyes wide open', in S.A. Goldsmith and L. Elizabeth (eds) *What We See: Advancing the Observations of Jane Jacobs*, Oakland, CA: New Village Press, xxi–xxvi.

Gottmann, J. (1961) *Megalopolis*, New York: Twentieth Century Fund.

Hall, P. and Pain, K. (eds) (2006) *The Polycentric Metropolis*, London: Earthscan.

Harris, R. (2011) 'The magpie and the bee: Jane Jacobs's magnificent obsession', in M. Page and T. Mennel (eds) *Reconsidering Jane Jacobs*, Washington, DC: American Planners Association, 65–82.

Harrison, J. and Hoyler, M. (eds) (2015) *Megaregions: Globalization's New Urban Form?*, Cheltenham: Edward Elgar.

Harvey, D. (1973) *Social Justice and the City*, London: Edward Arnold.

Harvey, D. (1982/1999) *The Limits to Capital*, Oxford: Blackwell.

Harvey, D. (1985) *The Urbanization of Capital*, Oxford: Blackwell.

Harvey, D. (2012) *Rebel Cities: From the Right to the City to the Urban Revolution*, London: Verso.

Harvey, D. (2014) *Seventeen Contradictions and the End of Capitalism*, London: Profile Books.

Harvey, D. (2015) 'The most dangerous book I have ever written', *Human Geography*, 8: 72–102.

Hassett, B. (2017) *Built on Bones: 15,000 Years of Urban Life and Death*, London: Bloomsbury Sigma.

Hodder, I. (2006) *Çatalhöyük: The Leopard's Tale*, London: Thames and Hudson.

Hodder, I. and Orton, C. (1976) *Spatial Analysis in Archaeology*, Cambridge, UK: Cambridge University Press.

Hirt, S. (ed) (2012a) *The Urban Wisdom of Jane Jacobs*, London: Routledge.

Hirt, S. (2012b) 'Jane Jacobs, modernity and knowledge', in S. Hirt (ed) *The Urban Wisdom of Jane Jacobs*, London: Routledge, 37–48.

Hutton, W. (2007) *The Writing on the Wall: China and the West in the 21st Century*, London: Abacus.

Ikeda, S. (2018) 'Cities, agriculture, and economic development: The debate over Jane Jacobs's "City-first thesis"', *Cosmos + Taxis*, 5: 3–4.

IPCC (Intergovernmental Panel on Climate Change) (2012) *Managing the Risks of Extreme Events and Disasters to Advance Climate Change Adaptation*, Cambridge: Cambridge University Press.

IPCC (2018) 'Summary for policymakers', in *Global warming of 1.5°C. An IPCC Special Report*, Geneva: Meteorological Organization. Available at: www.ipcc.ch/pdf/special-reports/sr15/sr15_spm_final.pdf

Jacobs, J. (1961) *The Death and Life of Great American Cities*, New York: Random House.

Jacobs, J. (1970) *The Economy of Cities*, New York: Vintage.

Jacobs, J. (1970/2016) '"The real problem of cities" (Speech at the Inaugural Earth Week Teach-in, Milwaukie, 1970', in S. Zipp and N. Storring (eds) *Vital Little Plans: The Short Works of Jane Jacobs*, New York: Random House, 198–223.

Jacobs, J. (1984) *Cities and the Wealth of Nations*, New York: Vintage.

Jacobs, J. (1989a) 'Systems of economic ethics, Part One', in F. Lawrence (ed) *Ethics in the Making: The Jane Jacobs Conference*, Atlanta, GA: Scholars Press, 211–50.

Jacobs, J. (1989b) 'Systems of economic ethics, Part Two', in F. Lawrence (ed) *Ethics in the Making: The Jane Jacobs Conference*, Atlanta, GA: Scholars Press, 251–86.

Jacobs, J. (1992) *Systems of Survival: A Dialogue on the Moral Foundations of Commerce and Politics*, New York: Vintage.

Jacobs, J. (1992/2016) '"Foreword" to *Death and Life of Great American Cities* (Modern Library Edition, 1992)', in S. Zipp and N. Storring (eds) *Vital Little Plans: The Short Works of Jane Jacobs*, New York: Random House, 276–84.

Jacobs, J. (1993/2016) '"Two ways to live" (Interview with David Warren, *The Idler*, 1993)', in S. Zipp and N. Storring (eds) *Vital Little Plans: The Short Works of Jane Jacobs*, New York: Random House, 285–325.

Jacobs, J. (1994a/2016) 'First letter to the Consumer Policy Institute (Energy Probe Research Foundation Newsletter, 1994)', in S. Zipp and N. Storring (eds) *Vital Little Plans: The Short Works of Jane Jacobs*, New York: Random House, 326–7.

Jacobs, J. (1994b/2016) 'Women as natural entrepreneurs', in S. Zipp and N. Storring (eds) *Vital Little Plans: The Short Works of Jane Jacobs*, New York: Random House, 328–38.

Jacobs, J. (2000) *The Nature of Economies*, New York: Vintage.

Jacobs, J. (2000/2016) '"Time and change as neighborhood allies" (Vincent Scully Prize Lecture, Washington, DC, 2000)', in S. Zipp and N. Storring (eds) *Vital Little Plans: The Short Works of Jane Jacobs*, New York: Random House, 352–63.

Jacobs, J. (2001a/2016) '"Canada's Hub Cities" (Speech at the C5 Conference, Winnipeg, 2001)', in S. Zipp and N. Storring (eds) *Vital Little Plans: The Short Works of Jane Jacobs*, New York: Random House, 363–69.

Jacobs, J. (2001b/2016) '"Efficiency and the Commons" (With Janice Gross Stein, "Grazing on the Commons" Conference, Toronto, 2001)', in S. Zipp and N. Storring (eds) *Vital Little Plans: The Short Works of Jane Jacobs*, New York: Random House, 370–80.

Jacobs, J. (2002/2016) '"The Sparrow Principle" (Excerpt from "Urban Economy and Development", interview with The World Bank, Toronto, 2002)', in S. Zipp and N. Storring (eds) *Vital Little Plans: The Short Works of Jane Jacobs*, New York: Random House, 381–405.

Jacobs, J. (2004) *Dark Age Ahead*, New York: Random House.

Jacobs, J. (2004a/2016) '"Uncovering the economy: A new hypothesis" (Excerpt from an unpublished work, 2004)', in S. Zipp and N. Storring (eds) *Vital Little Plans: The Short Works of Jane Jacobs*, New York: Random House, 406–31.

Jacobs, J. (2004b/2016) '"The end of the Plantation Age" (Lewis Mumford Lecture, New York, 2004)', in S. Zipp and N. Storring (eds) *Vital Little Plans: The Short Works of Jane Jacobs*, New York: Random House, 432–59.

Jacobs, J. (2005/2016) *The Last Interview and Other Conversations*, Brooklyn, NY: Melville House.

Jacques, M. (2009) *When China Rules the World*, London: Allen Lane.

Jonas, A. (1986) 'Book Review: Cities and the Wealth of Nations', *Progress in Human Geography*, 10: 131–3.

Jonas, A. and Ward, K. (2007) 'There's more than one way to be "serious" about city-regions', *International Journal of Urban and Regional Research*, 31: 647–56.

Joss, S. (2015) *Sustainable Cities: Governing for Urban Innovation*, London: Palgrave Macmillan.

Kaplan, J.O., Krumhardt, K.M., Ellis, E.C., Ruddiman, W.F., Lemmen, C. and Goldewijk, K.K. (2010) 'Holocene carbon emissions as a result of Anthropogenic land cover change', *The Holocene*, 21: 775–91.

Keeley, R.C. (1989) 'Some paths through Jane Jacobs's thought', in F. Lawrence (ed) *Ethics in the Making: The Jane Jacobs Conference*, Atlanta, GA: Scholars Press, 29–38.

Klein, N. (2014) *This Changes Everything*, London: Allen Lane.

Krugman, P. (1995) *Development, Geography and Economic Theory*, Cambridge, MA: MIT Press.

Laermans, R. (1993) 'Learning to consume: Early departmental stores and the shaping of modern consumer culture (1860–1914)', *Theory, Culture & Society*, 10: 79–102.

Lang, G. and Wunsch, M. (2009) *Genius of Common Sense: Jane Jacobs and the Story of The Death and Life of Great American Cities*, Boston, MA: David R. Godine.

Lang, R.E. (2003) *Edgeless Cities*, Washington, DC: Brookings Institution Press.

Laurence, P.L. (2011) 'The unknown Jane Jacobs: Geographer, propagandist, city planning idealist', in M. Page and T. Mennel (eds) *Reconsidering Jane Jacobs*, Washington, DC: American Planners Association, 15–36.

Laurence, P.L. (2016) *Becoming Jane Jacobs*, Philadelphia, PA: University of Pennsylvania Press.

Lawrence, F. (1989) *Ethics in Making a Living*, Atlanta, GA: Scholars Press.

Le Gates, R.T. and Stout, F. (eds) (2016) *The City Reader*, London: Routledge.

Leick, G. (2001) *Mesopotamia: The Invention of the City*, London: Penguin.

Lemmen, C. (2009) 'World distribution of land cover changes during pre- and protohistoric times and estimation of induced carbon releases', *Geoarchaeology: Human-Environment Connectivity*, 15: 303–12.

Lewis, S.L. and Maslin, M.A. (2015) 'Defining the Anthropocene', *Nature*, 519: 171–80.

Li, M. (2008) *The Rise of China and the Demise of the Capitalist World Economy*, New York: Monthly Review Press.

Lim, K.F. (2019) *On Shifting Foundations: State Rescaling, Policy Experimentation and Economic Restructuring in Post-1949 China*, Oxford: John Wiley.

Lin, G.C.S. (2002) 'The growth and structural change of Chinese cities: A contextual and geographic analysis', *Cities*, 19: 299–316.

Lin, G.C.S., Li, X., Yang F.F. and Fox, Z.Y. (2015) 'Strategizing urbanism in the era of neoliberalization: State power reshuffling, land development and municipal finance in urbanizing China', *Urban Studies*, 52: 1962–82.

Lucas, R. (1988) 'On the mechanics of economic development', *Journal of Monetary Economics*, 22: 3–42.

Mann, C.C. (2011) *1491: New Revelations of the Americas before Columbus*, New York: Vintage.

Massey, D. (1984) *Spatial Divisions of Labour: Social Structures and the Geography of Production*, London: Macmillan.

McHarg, I.L. (1969) *Design with Nature*, New York: John Wiley.

McMahon, D. (2018) *China's Great Wall of Debt*, London: Abacus.

Miller, T. (2012) *China's Urban Billion*, London: Zed Books.

Milman, O. (2018) 'Ex-Nasa scientist: 30 years on, world is failing "miserably" to address climate change', *The Guardian*, 19 June. Available at: www.theguardian.com/environment/2018/jun/19/james-hansen-nasa-scientist-climate-change-warning (accessed 22 October 2019).

Modelski, G. (2003) *World Cities: −3000 to 2000*, Washington, DC: Faros 2000.

Moore, J.W. (2014) 'The Capitalocene part I: On the nature and origins of our ecological crisis.' Available at: www.jasonwmoore.com/uploads/The_Capitalocene_Part_I_June_2014.pdf (accessed 10 January 2015).

Moore, J.W. (2015) *Capitalism and the Web of Life: Ecology and the Accumulation of Capital*, London: Verso.

Moore, J.W. (2016) *Anthropocene or Capitalocene? Nature, History, and the Crisis of Capitalism*, Oakland, CA: PM Press.

Morrone, F. (2017) 'The kind of problem gentrification is: The case of New York', *Cosmos + Taxis*, 4, 2–3.

Nelson, J.C. (2014) *Historical Atlas of the Eight Billion: World Population History 3000BCE to 2020*, Alexandra, VA: World History Maps Inc.

Neuwith, R. (2006) *Shadow Cities: A Billion Squatters, a New Urban World*, London: Routledge.

Nowlan, D. (1997) 'Jane Jacobs among the Economists', in M. Allen (ed) *Ideas that Matter: The Worlds of Jane Jacobs*, Owen Sound, ON: The Ginger Press, 111–13.

O'Brien, G. and O'Keefe, P. (2014) *Managing Adaptation to Climate Risk*, London: Routledge.

O'Brien, G., O'Keefe P. and Devisscher, T. (eds) (2011) *The Adaptation Continuum: Groundwork for the Future*, Saarbrucken, Germany: Lambert Academic Publishing.

O'Brien, G., O'Keefe, P. and Rose, J. (2007) 'Energy, poverty and governance', *International Journal of Environmental Studies*, 64(5): 607–18.

O'Keefe P., O'Brien, G. and Pearsall, N. (2010) *The Future of Energy Use*, London: Earthscan.

Owen, D. (2009) *Green Metropolis*, New York: Riverhead Books.

Packer, G. (2013) *The Unwinding: An Inner History of the New America*, New York: Faber & Faber.

Page, M. (2011) 'Introduction: More than meets the eye', in M. Page and T. Mennel (eds) *Reconsidering Jane Jacobs*, Washington, DC: American Planners Association, 3–14.

Patel, R. and Moore, J.W. (2018) *A History of the World in Seven Cheap Things*, London: Verso.

Pauketat, T.R. (2004) *Ancient Cahokia and the Mississippians*, Cambridge: Cambridge University Press.

Pauketat, T.R. (2007) *Chiefdoms and other Archaeological Delusions*, Lanham, MD: Altamira.

Pearce, F. (2010) *Peoplequake: Mass Migration, Ageing Nations and the Coming Population Crash*, London: Transworld.

Pearce, F. (2015) 'Myth of pristine Amazon rainforest busted as old cities reappear', *New Scientist*. Available at: www.newscientist.com/article/dn27945-myth-of-pristine-amazon-rainforest-busted-as-old-cities-reappear/ (accessed 15 October 2018).

Petrella, R. (1995) 'A global agora vs gated city-regions', *New Perspectives Quarterly*, Winter: 21–2.

Pieterse, E. (2008) *City Futures: Confronting the Crisis of Urban Development*, London: Zed Books.

Pirenne, H. (1925/1969) *Medieval Cities: Their Origins and the Revival of Trade*, Princeton, NJ: Princeton University Press.

Porter, M.E. (1998) 'Clusters and competition', in M.E. Porter (ed) *On Competition*, Cambridge, MA: HBS Press, 197–288.

Register, R. (2010) 'Jane Jacobs basics', in S.A. Goldsmith and L. Elizabeth (eds) *What We See: Advancing the Observations of Jane Jacobs*, Oakland, CA: New Village Press, 217–22.

Robinson, J., Scott, A.J. and Taylor, P.J. (2016) *Working/Housing: Urbanizing*, Berlin: Springer.

Rodseth, L., Wrangham, R., Harrigan, A. and Smuts, B.B. (1991) 'The human community as a primate society', *Current Anthropology*, 32: 221–54.

Romer, P. (1986) 'Increasing returns and long term growth', *Journal of Political Economy*, 94: 1002–37.

Ruddiman, W.F. (2003) 'Humans took control of greenhouse gases thousands of years ago', *Climate Change*, 61: 262–93.

Ruddiman, W.F. (2010) *Plows, Plagues, and Petroleum: How Humans Took Control of Climate*, Princeton, NJ: Princeton University Press.

Ruddiman, W.F. (2013) 'The Anthropocene', *Annual Review of Earth and Planetary Sciences*, 41: 1–24.

Ruddiman, W.F. and Ellis, E.C. (2009) 'Effect of per capita land use changes on Holocene forest clearance and CO_2 emissions', *Quaternary Science Reviews*, 28: 3011–15.

Sahlins, M. (2004) *Stone Age Economics*, London: Routledge.

Sardar, Z. (2010) 'Welcome to postnormal times', *Futures*, 42(5): 435–44. Available at: http://ziauddinsardar.com/2011/03/welcome-to-postnormal-times/ (accessed 22 October 2019).

Schumpeter, J. (1975) *Capitalism, Socialism and Democracy*, New York: Harper.

Scott, A.J. (2012) *A World in Emergence: Cities and Regions in the 21st Century*, Cheltenham: Edward Elgar.

Scott, A.J. and Soja, E.W. (1996) *The City: Los Angeles and Urban Theory at the End of the Twentieth Century*, Berkeley, CA: University of California Press.

Scott, A.J. and Storper, M. (2015) 'The nature of cities', *International Journal of Urban and Regional Research*, 39, 1–15.

Scott, J.C. (1998) *Seeing Like a State*, New Haven, CT: Yale University Press.

Shepard, W. (2015) *Ghost Cities of China*, London: Zed Books.

Smith, J., Karides, M., Becker, M., Brunelle, D., Chase-Dunn, C. and Porta, D.D. (2011) *Global Democracy and the World Social Forums*, Boulder, CO: Paradigm.

Smith, M.E., Ur, J. and Feinman, G.M. (2014) 'Jacobs' "cities first" model and archaeological reality', *International Journal of Urban and Regional Research*, 38, 1525–35.

Smith, M.L. (2003) *The Social Construction of Ancient Cities*, Washington, DC: Smithsonian Books.

Smith, N. and O'Keefe, P. (1980) 'Geography, Marx and the concept of nature', *Antipode*, 12: 30–9.

Soja, E. (2000) *Postmetropolis: Critical Studies of Cities and Regions*, Oxford: Blackwell.

Soja, E. (2010) 'Cities and states in geohistory', *Theory and Society*, 39: 361–76.

Srinivasan, B. (2017) *Americana: A 400-Year History of American Capitalism*, London: Penguin.

Steffen, W., Grinewald, J., Crutzen, P. and McNeill, J. (2011) 'The Anthropocene: Conceptual and historical perspectives', *Philosophical Transactions of the Royal Society A*, 369: 842–67.

Steel, C. (2008) *Hungry City: How Food Shapes Our Lives*, London: Chatto & Windus.

Szurmak, J. and Desrochers, P. (2017) 'Jane Jacobs as spontaneous economic order methodologist: Part 2 metaphors and methods', *Cosmos + Taxis*, 4: 21–48.

Taagepera, R. (1978) 'Size and duration of empires: Systematics of size', *Social Science Research*, 7: 108–27.

Taylor, P.J. (1996a) *The Way the Modern World Works: World Hegemony to World Impasse*, Chichester: John Wiley.

Taylor, P.J. (1996b) 'Embedded statism and the social sciences: opening up to new spaces', *Environment and Planning A*, 28: 1917–28.

Taylor, P.J. (2006) 'Cities within spaces of flows: Theses for a materialist understanding of the external relations of cities', in P.J. Taylor. B. Derudder, P. Saey and F. Witlox (eds) *Cities in Globalization: Practices, Policies and Theories*, London: Routledge, 287–97.

Taylor, P.J. (2011) 'Thesis on labour imperialism: How communist China used capitalist globalization to create the last great modern imperialism', *Political Geography*, 30(4): 175–7.

Taylor, P.J. (2012a) 'Extraordinary cities: Early "city-ness" and the origins of agriculture and states', *International Journal of Urban and Regional Research*, 36: 415–47.

Taylor, P.J. (2012b) 'Transition towns and world cities: Towards green networks of cities', *Local Environment*, 17: 495–508.

Taylor, P.J. (2013) *Extraordinary Cities: Millennia of Moral Syndromes, World-Systems and City/State Relations*, Cheltenham: Edward Elgar.

Taylor, P.J. (2015) 'Post-Childe, post-Wirth: Response to Smith, Ur and Feinman', *International Journal of Urban and Regional Research*, 39: 168–71.

Taylor, P.J. (2016) 'Corporate Social Science and the Loss of Curiosity', *Items*, 2 August (New York: Social Science Research Council). Available at: http://items.ssrc.org/corporate-social-science-and-the-loss-of-curiosity/ (accessed 22 October 2019).

Taylor, P.J. (2017a) 'Cities in climate change', *International Journal of Urban Sciences*, 21(1): 1–14.

Taylor, P.J. (2017b) 'Incorporating geographical imagination into early demographic estimates', Conference Paper, *Network Evolutions* at UrbNet, Aarhus University, January.

Taylor, P.J. (2017c) 'The new political geography of corporate globalization', *L'espace politique*, 32: 2017–22.

Taylor, P.J. and Derudder, B. (2016) *World City Network: A Global Urban Analysis* (2nd edn), London: Routledge.

Taylor, P.J., O'Brien, G. and O'Keefe, P. (2015) 'Commentary: Human control of climate: Introducing cities', *Environment and Planning A*, 47: 1023–8.

Taylor, P.J., O'Brien, G. and O'Keefe, P. (2016) 'Eleven antitheses on cities and states: Challenging the mindscape of chronology and chorography in Anthropogenic climate change', *ACME: An International Journal for Critical Geographies*, 15(2): 393–417.

Taylor, P.J., Ni, F., Derudder, B., Hoyler, M., Huang, J. and Witlox, F. (eds) (2010) *Global Urban Analysis: A Survey of Cities in Globalization*, London: Earthscan.

Thiel, P. (2014) *Zero to One: Notes on Startups, or How to Build the Future*, New York: Crown Business.

UNEP (United Nations Environment Programme) (2018) *The Emissions Gap Report*, Nairobi: UNEP. Available at: www.unenvironment.org/emissionsgap (accessed 22 October 2019).

Urban Theory Lab-GSD (2014) 'Visualising an urbanized planet – Materials', in N. Brenner (ed) *Implosions/Explosions: Towards a Study of Planetary Urbanization*, Berlin: Jovis, 460–75.

van den Berg, M. (2018) 'The discursive uses of Jane Jacobs for the genderfying city: Understanding the productions of space for post-Fordist gender notions', *Urban Studies*, 55: 751–66.

Verbruggen, R., Hoyler, M. and Taylor P.J. (2014) 'The networked city', in P. Knox (ed) *Atlas of Cities*, Princeton, NJ: Princeton University Press, 34–51.

Vernon, R. (1966) 'International investment and international trade in the product cycle', *Quarterly Journal of Economics*, 80(2): 190–207.

von Thünen, J.H. (1826/1966) *Isolated State*, New York: Pergamon Press.

Wallerstein, I. (1974) *The Modern World-System: Capitalist Agriculture and the Origins of the European World-Economy in the Sixteenth Century*, New York: Academic Press.

Wallerstein, I. (1979) *The Capitalist World-Economy*, Cambridge, UK: Cambridge University Press.

Wallerstein, I. (1984) *The Politics of the World-Economy*, Cambridge: Cambridge University Press.

Wallerstein, I. (1991) *Unthinking Social Science: The Limits of Nineteenth Century Paradigms*, Cambridge: Polity Press.

Wallerstein, I. (1992) 'The West, capitalism and the modern world-system', *Review*, 15: 561–619.

Wallerstein, I. (2004) *World-Systems Analysis*, Durham, NC: Duke University Press.

Wallerstein, I., Juma, C., Keller, E.F., Kocka, J., Lecourt, D., Mudimbe, Y.Y., et al (1996) *Open the Social Sciences: Report of the Gulbenkian Commission on the Restructuring of the Social Sciences*, Stanford, CA: Stanford University Press.

Weaver, W. (1948) 'Science and complexity', *American Scientist*, 36: 536–44.

Weber, A. (1899) *The Growth of Cities in the Nineteenth Century: A Study in Statistics*, Ithaca, NY: Cornell University Press.

Weber, M. (1921/1958) *The City*, New York: Free Press.

Wenz, P.S. (1995) 'Book review: *Design with Nature, Ecotecture*', *The Journal of Ecological Design*. Available at: www.ecotecture.com/reviews/mcharg2.html (accessed 22 October 2019).

Wheatley, P. (1969) *The Pivot of the Four Quarters*, Chicago, IL: University of Chicago Press.

Wiedmann, T.O., Schand, H., Lenzen, M., Moranc, D., et al (2015) 'The material footprint of nations', *PNAS*, 112(20): 6271–6.

Williams, R. (1973) *The Country and the City*, Oxford: Oxford University Press.

Yoffee, N. (2005) *Myths of the Archaic State: Evolution of the Earliest Cities, States, and Civilizations*, Cambridge: Cambridge University Press.

Zhang, L.-Y. (2003) 'Economic development in Shanghai and the role of the state', *Urban Studies*, 40: 1549–72.

Zipp, S. and Storring, N. (eds) (2016a) *Vital Little Plans: The Short Works of Jane Jacobs*, New York: Random House.

Zipp, S. and Storring, N. (2016b) 'Introduction', in S. Zipp and N. Storring (eds) *Vital Little Plans: The Short Works of Jane Jacobs*, New York: Random House, xvi–xxxvi.

Appendix: Primer on Climate Change Policy

Many believe that climate change is the greatest threat we face. But there are others who do not. Although the international community has put in place a system for tackling climate change, it is not making sufficient progress to avoid a global temperature rise that could lead to dangerous climate events. In addition, although there is increasing evidence of adverse events being driven by climate change, some nations such as the USA do not support actions, and the energy producers are clearly using their immense political influence to limit efforts being made to reduce conventional energy use. This Primer discusses climate change and posits that cities are best placed to lead on reducing greenhouse gas emissions as well as the adaptations needed to deal with changes driven by climate change.

Climate change is a natural phenomenon that depends on relationships between the Earth, Moon, Sun and other planets in our solar system and the composition of the Earth's atmosphere. The amount of what are termed 'greenhouse gases', such as carbon dioxide (CO_2), determine how much energy from the sun is trapped in the atmosphere. The more greenhouses gases present, the more heat is trapped, and the warmer the atmosphere becomes. This is called the 'greenhouse effect', a natural phenomenon that fluctuates cyclically. Over time the Earth has experienced both ice ages and warm periods, with the last ice age ending some 12,000 years ago.

With the advent of the Industrial Revolution, increasing use of fossil fuels massively increased the amount of greenhouse gases, particularly CO_2, emitted into the atmosphere. This has grown from a concentration of some 280 parts per million (ppm) prior to the Industrial Revolution to some 400ppm today. And it is still growing. Since the Industrial Revolution the global average temperature has risen by 1 degree Celsius. Climate scientists believe that a rise of more than 1.5 degrees Celsius will lead to dangerous climate change.

What has been the response? The international community has recognised the dangers of climate change. In 1994 the United Nations Framework Convention on Climate Change (UNFCCC) came into force. Its role is to find the means of reducing greenhouse gas emissions. It initially set a target to reduce emission rates to 1990 levels by 2000. This proved to be completely unrealistic, and in 1997 UNFCCC established the Kyoto Protocol. This had a very modest target of just over 5 per cent reduction in emissions along with a flexible approach to meeting targets. Although the protocol had sufficient signatories to make it binding, the world's largest emitter, the USA, refused to sign, claiming it might damage its economy. The Kyoto Protocol did meet its modest targets, but some claim that emission reductions were more driven by the economic recession of 2008 than actual efforts to reduce emissions.

The international community has also recognised the need for funding greenhouse gas reductions (mitigation) and adjustments needed, driven by climate change (adaptation). At the UNFCCC 2009 meeting in Copenhagen it was agreed to establish the Green Climate Fund to assist poorer countries in paying for their mitigation and adaptation policies. In Berlin at a pledging conference in 2014 some 30 countries announced their intention to contribute to the Fund. Since then, many of these nations have increased their pledges and more have joined the fund, including China, which pledged to give US$3.1 billion over a non-specific time frame. A total amount of US$10.3 billion was legally pledged and signed to be paid into the fund by 43 countries by 2018. In 2017 US$4.6 billion had already been paid, with Japan the biggest contributor, followed by the USA, UK, Germany, France and Sweden. However, the USA has since refused to pay the additional US$2 billion it had committed. President Donald Trump argues that the Paris Climate Agreement is flawed and will damage the USA; the USA has now indicated that it will leave the Treaty in 2020. It is uncertain if the other contributors will make up the balance of the Green Climate Fund, and it is also unclear where the rest of the funds will come from. It is also believed that Trump hostility to the Paris Agreement will encourage others to backtrack on commitments (Curtin, 2018).

The Paris Agreement was a successor to the Kyoto Protocol, UNFCCC's next push to reduce greenhouse gases. There are some positive aspects to the Agreement, such as the need to minimise and address loss and damage, the need for more early warning systems, the need for cities, regions and local authorities to scale up efforts to reduce emissions and build resilience, and the need for funds to support climate action in developing countries. Signatories to the Agreement are required to specify greenhouse gas reductions through nationally determined contributions

(NDCs). The levels specified are voluntary. The commitments made to date are insufficient to meet the goal of keeping the average global temperature increase below 1.5 degrees Celsius (UNEP, 2018). According to Climate Transparency (2018), 15 of the G20 nations have since reported an actual rise in CO_2 emissions. Further, there has been an increase of 50 per cent in subsidies to the fossil fuel sector. Some US$147 billion in subsidies was paid by the G20 nations in 2016, despite their pledge, made more than 10 years ago, to phase them out.

The Intergovernmental Panel on Climate Change (IPCC) believes that we are likely to have an increase of 1.5 degrees Celsius of warming between 2030 and 2052. Although there are serious adverse effects associated with such a temperature rise, many states still seem unwilling to act. At the recent annual meeting of UNFCCC in Katowice in Poland, there was a reluctance to make any meaningful progress on increasing emission targets in the Paris Agreement. It is easy to understand the reluctance – many states have high energy use. This is associated with standards of living – the greater the access to energy, the higher the standard of living. Reducing emissions implies reducing energy use that could have an impact on standards of living. Few politicians are willing to risk that. Further, the energy sector is politically influential and lobbies states to retain the current energy mix.

Of course there are alternatives such as renewables and nuclear power. As with any energy system, however, the alternatives have problems. Renewables require a different and more diverse approach. In many countries this could weaken the position of fossil fuel organisations. Given the economic power of the current energy sector, it is unlikely that they would willingly support such changes. Nuclear power is costly, and the industry has not solved the waste problem.

Where does this leave us? Well, it seems very likely that things will drift on until something drastic does happen. There is nothing like a disaster to stimulate politicians to respond to changes that the public would want. Although we have been experiencing some quite severe weather events, they are not, as yet, sufficiently severe to generate such a reaction. But this is not completely the case. Some parts of the world are experiencing dramatic change such as Inuit communities in the Arctic, where temperatures are rising rapidly, affecting ice conditions, and a number of islands in different parts of the world, such as Tuvalu in the Pacific, are low-lying and vulnerable to rising sea levels. However, the prevailing view of how we react to climate change is summed up in the Giddens' Paradox, which is stated as follows: 'since the dangers posed by global warming aren't tangible, immediate or visible in the course of day-to-day life – however awesome they may appear – many will sit on their hands and do nothing concrete about them. Yet waiting until they become visible and acute before being stirred to serious action will, by definition, be too late' (Giddens, 2009: 2).

This implies that people believe that climate science lacks certainty, and without certainty policy-makers will not act. Many may be sympathetic to this view; indeed, it takes someone of the status of David Attenborough to make people aware of the damage caused to the marine environment by plastic waste. Although some action has been taken against plastic waste, it is too little when compared to the overall scale of the problem. For instance, the ecological damage caused by palm oil plantations and the deforestation caused by beef production are conspicuous by their absence from mainstream discourse. The basic reality is that the developed world is caught in a model that defines personal status through consumption; the fancier the car, house and other stuff you have defines both your place in, and value to, society. In addition, leaders of the developing and industrialising worlds aspire to the Western standard of living. That is entirely understandable. The media, dominated by the West, promotes that conspicuous consumptive lifestyles are the choice of leading citizens. At the same time there is a genuine desire to lift people out of poverty. However, it is often not realised that many aspects of life in the developed world can be one of the causes of poverty in the developing world. Thus, we argue that Giddens is wrong: the causes of the unwillingness to act on climate change are much deeper than the tangibility of the dangers.

The state is the principal unit of government and it is the assemblage of states that work together (or try to) to deliver agreed international aims. They are torn by what they perceive as national interests and how these fit into international targets. It is often a contradiction. Which politician will argue for higher fossil fuel costs in order to save the planet? That would be perceived as lowering living standards. Would many politicians argue the case? Few, if any. In a recent interview, James Hansen, a former NASA scientist, who has warned for some 30 years of the dangers of climate change, believes that states are the problem. Hansen believes that many national leaders are only pretending to solve the problem. He argues that the main driver is the domination of the energy lobby and the need for ongoing growth (Milman, 2018).

Running parallel with UN climate programmes have been global environmental initiatives that do not necessarily involve national governments. At the Rio 'Earth Summit' in 1992, the Agenda 21 agreement encouraged local policy development and UN-Habitat started its Sustainable Cities Programme at about the same time. With the faltering of intergovernmental negotiations, cities have come to be seen as an alternative route for responding to climate change. This has been particularly the case in the USA after President Trump's repudiation of the Paris Climate Agreement.

Why cities? Quite simply there is recognition that innovation and change is required and cities are the places especially associated with innovation and change. Furthermore there is the sheer scale of contemporary urbanisation. It is likely that the urban population will double in size by 2050. The demand for resources, including energy, is therefore likely to increase. Hence it seems logical that cities should lead the drive on climate mitigation and adaptation. Cities are agile, connected to their citizens, able to implement appropriate solutions to the problems they face and through networks can exchange solutions they have developed. There is clear evidence that people who live in cities, particularly compact cities, have a lower environmental impact than citizens who live in more rural areas, particularly in terms of energy use (Owen, 2009).

A number of cities are working to address climate change. However, the reality is that generally cities lack the power to make effective and far-reaching decisions. We have seen a drive by national governments to centralise power at the expense of municipalities' ability to act at the local level. Increasingly local government is becoming merely the manager of central government policy. But for cities to deal with climate change, decentralisation is needed. This process can act at the municipal level as well as regional and national levels.

Here are some common initiatives:

- Shift from outsourcing to direct provision by a public authority. In many parts of the world this has worked successfully with water and sanitation services.
- Re-municipalisation has also been to be extended to cover energy use. This gives a much clearer focus on mitigation. How can re-municipalisation of energy help with mitigation? Public ownership of the energy network means that municipalities can focus on reducing conventional energy use and promoting alternatives. Buildings alone consume some 40 per cent of energy. There is a need to make buildings produce energy. Good design can minimise the amount of energy a building requires.
- A further major energy user is transport. Reducing transport energy use requires a number of options. First is the development of effective public transport such as metro systems and efficient buses. Second, autonomous electric vehicles can replace motor vehicles. Studies suggest that one autonomous vehicle can replace fourteen conventional vehicles. Third, we need to ensure that cities are compact and that walking routes are both plentiful and well designed.

- Provisioning cities makes a contribution to climate change; for example, the meat-producing sector generates a similar amount of greenhouse gases as the transport sector. Meat eating is increasing rapidly across the world and there are concerns about the environmental impacts this will have. In the future the amount of land needed for agriculture will rise sharply. Building vertical farms – stacked layers of food production – within the city would considerably lower its environmental impact.

As well as mitigating the effects of climate change through the city there is also a need to ensure that adverse consequences driven by climate change can be coped with. This is a major challenge. Many existing cities are vulnerable to adverse climate impacts. For example, some cities have developed in places likely to flood. It is important that cities identify the risks they are likely to face. This will allow the development of measures needed to cope with adverse climate impacts. Measures are classified as hard or soft. Hard measures can include flood defences and structures to provide shade. Soft measures include building codes and education. The problems that adaptation measures face can be seen in two perspectives. The first is dealing with existing built environments. The second is putting in place measures that will ensure that future built environments are developed to be ready to face adverse climate events.

The reality is that most new development will take place in Asia and Africa. Governments are unlikely to let cities re-municipalise. New build will be primarily developer-led and it is highly likely that there will little positive innovation in city development. Economic approaches will remain based on conspicuous consumption. There will be some cities in Europe that will continue to re-municipalise, but the impact they will make on climate change will be small. The demand for energy to deal with growing populations will continue to rise. Countries with large supplies of fossil fuels, such as India, will use those fuels to stimulate economic and physical development. It seems the real potential of cities as centres of innovation will be inconsistent, marginal and therefore wholly inadequate as a means of tackling global climate change.

To sum up: in confronting anthropogenic climate change it seems that humans have been quite good at international institution building, and some of this has transferred to cities, but the resulting institutions are found to be not up to the job. Why, and where does this leave us? The evidence that we are interfering with the climate system is increasingly convincing. We do need to improve climate science and efforts to do this are being made. We do have ways of engaging with climate change such as renewable technologies, efficiency improvements and adaptation measures. We have an international agreement to reduce greenhouse gas concentrations to a safe level. We have established a fund to help poorer nations cope with climate change. It would seem that all is in place to successfully tackle this critical problem. But the reality is very different.

States are unlikely to act as they have conflicting interests. For example, although renewable technologies are readily available, governments often seem reluctant to support them. The reality is that fossil fuels remain heavily subsidised. Fossil fuel companies are also able to lobby far more effectively than the manufacturers of renewable technologies. It has become apparent that cities offer a real opportunity to address climate change, but it is unlikely that central governments will give them the powers and funding to act. The centralising tendency is overpowering. And we can see that although there is recognition that the developing world needs assistance to address climate problems, the mechanism to address this, the Green Climate Fund, is seriously underfunded. Perhaps the key issue is the amount of urban development that will happen in the coming decades; this will need to ensure it takes into account a form of adaptation that might work. Is this likely? Probably not. In essence, the response to climate change is seriously weak.

And where precisely are we now? We come back to the 2018 IPCC Report that says carbon emissions are continuing to rise. The IPCC believes that if we do not reduce carbon emissions by 2030, we will enter an era of dangerous climatic effects. In other words, we are in a climate emergency.

Index

References to tables and figures are shown in *italics*

A
Abu-Lughod, J. 40
academia, separateness of 2–3
Academy of Social Sciences 10
adaptation spectrum 117–18, *118*
Advertising-Big Data-Social Media complex 15, 90, 100, 122
Age of Human Capital 29–30, 36
Agenda 21 agreement 140–1
agglomerations 38, 66–7, 75, 89
agriculture
    and anthropogenic climate change 61–2, 78–80, 81
    cheap food 115
    and cities 30, 37, 52–9, 79–80, 81–2, 93
    and diffusion 55, 56
    irrigation 81
    land use model 54
    and migration 55, 56
    original plant domestication 53–5
    Plantation Age 28–9, 36, 40–1, 115
    and population growth 79
    subsistence 58–9, 62, 86
    transmission of by cities 55–7
    vertical farms 142
Amazonia 54–5
Amin, A. 60
Amin, S. 40
Angus, I. 85
Anthropocene 91, 92, 93, 112
anthropogenic climate change
    adaptation spectrum 117–18, *118*
    and agriculture 61–2, 78–80, 81
    chronology of 61–4, 76–84, *80*, 85–7, 93
    and cities 61–6, 79–84, 93
    as doubly complex 3–4, 6–8
    early climate change thesis 61–3, 76–80, 85–7
    as existential threat 4–5
    and international relations 5–6
    international response to 138–43
    mitigation policies 7, 96, 117–18, *118*, 138, 141
    new chronology 61–2, 76–80, 85–7, 90–3
    primer on climate change policy 137–43
    and scientific input 6–8
    slow/fast division 61–2, 83
    and social science 6–10, 7
    and stewardship 97, 118–22
    trans-modern interventions 84–93
    trans-modern narrative on 80–4
archaeology 30, 39, 50, 53, 56, 79, 81, 85
Askins, K. 42, 88
Attenborough, David 1, 7

B
Belich, J. 60, 83
Bookchin, M. 89
Boserup, E. 79, 82
bottom-up process 13, 33, 42, 67, 70, 73, 88, 102–3, 120
Brand, S. 75
Braudel, F. 40, 49, 53, 70
breakdown of civilization 101, 102

Brenner, N. 60, 87
Britain 8, 10, 67, 106, 108, 116
Brooke, J.L. 60, 83
Bulkeley, H. 73

C

Campanella, T.J. 107, 118
Canada 26
capitalism 40, 41, 57–8, 63, 83–4, 90, 92–3
    characteristics of expansion 113–16
    superseding 111–16
    see also economic growth; urban demand
Capitalocene 93, 112
carbon dioxide (CO_2) 77, 78, 81, 82, 83, 137, 139
carbon emissions 3–4, 8, 62, 63–4, 139
carboniferous capitalism 63
care work 114
Carter, H. 85
Castells, M. 39, 41, 49, 52
Castree, N. 70
Çatalhöyük, Anatolia 53, 55, 74
cheap care 114
cheap energy 115
cheap food 115
cheap lives 116
cheap money 113–14
cheap nature 113
cheap things and capitalism 113–16
cheap work 114
Childe, V.G. 53, 60, 85
China
    de-urbanisation 119
    in modern world-system 83
    planetary stewardship 120–2
    rapid urban growth 104–7
    urban demand process 107–11
Chinese Communist Party (CCP) 120–1
Cichello, A. 20
cities
    adaptation spectrum 118, *118*
    and agriculture 30, 37, 52–9, 79–80, 81–2, 93
    and anthropogenic climate change 61–6, 79–84, 93
    in China 104–11, 120–2
    cities first thesis 30, 39, 53, 59, 62, 78, 79–80, 85–6
    communication advantages of 74–6
    and creation of states 50–1
    creativity 76, 103–4, 119, 120
    and economic development 22–6, 31–3, 37–8, 41–2, 58, 67, 72–6, 82–3, 88–9, 103
    green 65–6, 89
    and industrialisation 57–8
    initiatives addressing climate change 140–2
    and macro-social change 48–52
    and nature 42, 68–9
    as new polycentric structures 66–7
    occidental and oriental 108
    poor cities 40
    population of 60, 75, 79, 81, 82, 83, 86–7
    as process 39, 42, 52, 60, 65, 66, 67, 69, 73, 89, 90, 119
    reinventing as stewardship 118–22
    and rural places 59–61
    as social development 51–2
    trans-modern interventions 84–93
    see also urban demand
cities first thesis 39, 53, 59, 62, 78, 79–81, 85–6
city net-prints 64
city–state relations 58, 67, 82–3
city-states 50–1, 82, 103
civilisation breakdowns 101, 102
class politics 31–2
Clement, C.R. 55, 83
Climate Transparency 139
commercial autonomy 108–10
Commercial moral syndrome 21, 30, 32, 40, 49, 103
communication advantages of cities 74–6
Communist Party, China 120–1
consumerism/consumption *see* urban demand
corporate power 99–102

corporate social science 99–100
creative destructions 42, 49
critical research agendas 85–91
critical unthinking 10–14
Cronon, W. 60
Crosby, A.W. 57
Crutzen, P. 92

D
de-industrialisation 40, 58
decentralisation 109, 120–1, 141
demand *see* urban demand
Deng Xiaoping 110
Desrochers, P. 31
development of underdevelopment 40
Diamond, J. 28, 55
disease pandemics 82, 83
dynamic stability/systems 34–5, 42, 84, 121

E
ecological imperialism 57
ecological planning 69
economic development
   and anthropogenic climate change 72–6, 80
   and Chinese cities 104–10, 120
   and cities 22–6, 31–3, 37–8, 41–2, 58, 67, 72–6, 82–3, 88–9, 103
   exports 23–4, *24*, 34, 73, 107–9, 109, 110
   imports 23–5, *24*, 34, 36, 37–8, 67, 73, 76, 79, 107–8, 109, 110
   and nature 34–6
   new work 23–5, *24*, 29, 31–3, 76, 103, 107–8, 110, 118, 119
   old work 23–5, *24*, 29, 31, 33, 36, 41, 76, 103, 119
   poor cities 40
   stagnation 25, 32, 33, 36, 59, 119
   see also capitalism; supply issues; urban demand
ecosystems 34–5, 68, 78
edgeless cities 67
Ehrlich, P.R. 75
energy production 64, 115, 139, 141, 142

entrepreneurship 24–5, 33
everyday urbanism 88
evolution, change as 50
existential threat 4–5
exports 23–4, *24*, 34, 73, 107–9, 109, 110

F
farming *see* agriculture
Fenby, J. 110
food industry 115
   see also agriculture
foragers 28–9, 30
Foresight report 73
fossil fuels 137, 139, 140, 142, 143
Frank, A.G. 40
Froy, F. 23, 26
Funtowicz, S. 40

G
Gamble, C. 50, 81
gentrification 31
geo-engineering 117
geological time periods 91–3
geopolitics 49
ghost cities 109
Gibson-Graham, J.K. 41
Giddens' Paradox 139–40
Gignoux, C.R. 79
glacial cycles 77
Glaeser, E. 26, 30, 74, 75, 103–4
Global North/South divide 60, 90, 107
Goldsmith, S.A. 17
Gottmann, J. 66
green cities 65–6, 89
Green Climate Fund 138, 143
greenhouse gases (carbon dioxide and methane)
   changing levels 61, 62, 63, 77–8, *80*, 81, 82, 83, 137
   and cheap energy 115
   and food industry 115
   and greenhouse effect 137
   international response 138–9
Gregg, S.A. 86
Guardian moral syndrome 21, 29, 32–3, 40, 41, 49, 103

H

habitat destruction 35–6, 84
Hansen, James 140
Harris, R. 20, 27
Harvey, D. 51, 85, 89, 90, 112
Hassett, B. ix, 56
Hirt, S. 28
Holocene 91–2, 93
hunter-gatherers 37, 53, 56, 74, 79, 84

I

ice ages 61, 62, 77, 91, 93, 137
imports 23–5, 24, 34, 36, 37–8, 67, 73, 76, 79, 107–8, 109, 110
Industrial Revolution 29, 40–1, 47, 47, 57–8, 61, 62, 64, 92, 137
innovation 23, 25, 32, 33, 67, 76, 89, 103–4, 141, 142
inter-city relations 25–6
Intergovernmental Panel on Climate Change (IPCC) 7, 42, 46, 47, 68, 97, 139, 143
International Commission on Stratigraphy 91, 92
International Geologic Time Scale 91
international relations, challenges of 5–6, 8, 64
International Union of Geological Sciences 91

J

Jacobs, Jane 12–13, 17–43
  and bottom-up process and action 13, 42, 120
  cities first thesis 30, 39, 53, 59, 62, 78, 79–80, 85–6
  and climate change 18
  and economics 12, 15, 22–7, 40–1, 49, 52, 59, 72–6, 89, 97–8, 102–3, 107–10, 119, 121
  and habitat destruction 84
  and history 27–30
  intersecting with other thinkers 38–43
  and knowledge building 19–22
  legacy of 17–18
  and nature 33–6, 42, 69, 71
  Plantation Age 28–9, 36, 40–1, 115
  and politics 30–3, 49
  and unthinking 12–14
Japan 104

K

kairos 41, 100–1, 106
Kaplan, J.O. 77
Keeley, R.C. 20
Klein, N. 70, 95
knowledge
  bottom-up 42
  building 19–22, 41, 42
  knowledge–action gap 70
  knowledge flows in cities 75–6
Krugman, P. 26, 30
Kyoto Protocol 18, 138

L

land clearance 62, 80, 85–6, 93
land use model 54, 55, 87
Lang, R.E. 67
Laurence, P.L. 18, 19, 28
Leick, G. 82
Lemmen, C. 85–6
Lim, K.F. 120
Lin, G.C.S. 108
local economic growth theory 26
Los Angeles 15, 37, 106, 118
Lucas, R. 26
Lynne, E. 17

M

macro-social change, sources of 48–52, 90
Manchester 15, 25, 29
Mann 59, 83
manufacturing 29, 47, 114
markets 13, 32, 33, 38, 40, 103
Marshall, Alfred 26–7
Marxism 51, 57, 89, 112
mass extinctions 5
mass production 29, 41
Massey, D. 42
McHarg, I.L. 69

McMahon, D. 109, 110, 120
meat production 142
megalopolis 66
Mesopotamia 30, 53, 67, 74, 79, 80, 81, 85, 103
methane (CH$_4$) 61, 62, 77, 78, *80*, 82, 83
migration to cities 60, 98, 105, 108
Milanković, Milutin 91
mitigation policies 7, 96, 117–18, *118*, 138, 141
modern world-system 82–3, 92–3, 100–2, 106–7
money 113–14
monocultures 29
Moore, J. 57, 68, 72, 92, 112–16
moral codes 21
Morrone, F. 31

N

nature
    anthropogenic climate change 77–8
    cheap nature 113
    and cities 42, 68–9
    and Jacobs, Jane 33–6, 42, 69, 71
    'nature in control' 77–8
    stewardship 119
    and Urbanocene 95
Nelson, J.C. 86
Neolithic populations 79, 85–6
Netherlands 106
Neuwirth, R. 67
new work 23–5, *24*, 29, 31–3, 76, 103, 107–8, 110, 118, 119
non-local environments/resources 36, 81, 84, 103

O

oceans 64
old work 23–5, *24*, 29, 31, 33, 36, 41, 76, 103, 119
orbital changes of Earth 77
organised complexity 4, 19–20, 33–4, 35, 43, 73
Owen, D. 42

P

Packer, G. 41–2
Page, M. 30
Pain, R. 42, 88
Paris Agreement 2015 4, 5, 6, 7, 138–9
participatory action research 42, 88
Patel, R. 113–16
Pauketat, T.R. 39, 50
Pearce, F. 75
Pieterse, E. 88
Pirenne, H. 27, 40
planetary urbanisation 60, 64, 66, 87, 89, 107
Planetizen survey 17
planning 22, 69
Plantation Age 28–9, 36, 40–1, 115
policy-making
    adaptation spectrum 117–18, *118*
    and bottom-up actions 88
    in Chinese cities 108–10
    and existential threat 4–5
    international 137–42
    and mainstream science 6–7, 46–7, *47*, 96
    mitigation policies 7, 96, 117–18, *118*, 138, 141
    and nature of cities 61–9
    policy dilemmas 95
    simple and complex problems 3–4
    states' ability 5–10, 23
    see also supply issues; urban demand
politics
    class politics 31–2
    making fundamental changes 111–16
    party politics 41–2, 101
    politics of change through cities 88–90
    populism 42, 101–2
    radical 90, 111–16
polycentric city-regions 66–7
population growth 60, 75, 79, 81, 82, 86–7
populism 42, 101–2
posterity cities 118, 119, 121

R

radical dependency school 40
radical politics 111–16
Ravetz, J.R. 40

re-municipalisation 141, 142
'release from proximity' 81
renewable energy 139
research agendas 85–91, 99–100
research methods 19, 42, 88
resilient cities 42, 49, 81, 118
Rodseth, L. 81
Romer, P. 26
Ruddiman, W.F. 61–2, 73, 76–80, *80*, 81, 82–3, 85–6, 87, 93
rural places 59–61
rural–urban migration 60, 98, 105, 108

S
Sahlins, M. 56, 79, 84
Schumpeter, J. 49
Scott, A. 89
Scott, J. 33, 40–1, 67
Shepard, W. 120
simple problems 3
slave trade 116
Smith, Adam 22–3, 35
Smith, M.L. 50
Smith, N. 113
social knowledge, creation of 10–12
social reproduction 80–2, 84, 114
social science(s)
   corporate social science 99–100
   definition of 98–9
   and new social knowledge 10–12
   role of 6–8, 7, 96
   state-centrism of 8–10, 99
   university social science 99
   and urban demand 98–9
social space 49
Soja, E. 30, 80, *80*
solar radiation 77, 78
spaces of flows and place 49, 51
Srinivasan, B. 116
stagnation 25, 32, 33, 36, 59, 119
states
   challenges of international relations 5–6, 101–2
   and complexity of problem 63–4
   conflicting interests 143
   de-linking from world economy 40
   and economics 23

energy production 64
and industrialisation 57–8
instrumentalist conception of 46, 63
interstate relations 49, 64
and macro-social change 48–51
origins of 50–1, 82–3
as the problem 140
and social sciences 8–10, 41
state-centric thinking 8–10, 41, 57, 99
stewardship 97, 118–22
Stewardship-Big Data-Social Media complex 122
stone age economics 56
Storring, N. 17, 18, 22, 31, 32, 33, 34, 36, 71
subsistence agriculture 58–9, 62, 86
supply issues 26
   Chinese cities 107–8
   and cities first thesis 29, 30, 39, 53, 59, 62, 78, 79–81, 85–6
   and climate change policy 46–7, 47, 73, 88.97
   and economic development 37–8
   and energy production 64
   and poorer cities 26
   prioritising of 46–7, 51, 58, 64, 73, 87, 88, 97, 112
   see also economic development; urban demand
sustainable cities 66, 89, 118, 140–1
Szurmak, J. 31

T
taxation 33, 117
technology, and solutions to climate change 65–6
temperatures *80*, 137, 139
thesis/antithesis format 48
Thiel, P. 40
'thing theory' of development 34, 119
Thrift, N. 60
trade unions 112, 114
transport systems 141
Trump, Donald 138

U
United Nations
   challenges of international relations 5–6

INDEX

UN Climate Change Conferences (COP) 46, 87
UN Framework Convention on Climate Change (UNFCCC) 5, 138–9
UN Intergovernmental Panel on Climate Change (IPCC) 7, 42, 46, 47, 68, 97, 139, 143
 and urban demand 98
United States
 city economic growth 25
 climate change policy 138
 consumption cities 122
 Kyoto Protocol 138
 Los Angeles 15, 37, 106, 118
 rapid urban growth 106
 slave trade 116
universities, role of 11, 99
unthinking 2, 10–14, 45
urban demand 95–122
 and agriculture 27–9, 30, 52–7, 59, 60, 62, 79–82, 86, 93
 anthropogenic climate change 14, 15, 61–6, 72–3, 76, 79–84, 86–8, 93
 in China 104–11
 cities first thesis 30, 39, 53, 59, 62, 78, 79–80, 85–6
 concept of 15–16
 consumerism 15, 46–7, 47, 65–6, 88–9, 140
 contemporary knowledge of 97–104
 and Jacobs, Jane 22–7, 36–8, 102–4, 107–8
 at local level 67, 108–10
 and mainstream climate science 46–7, 47
 and macro-social change 47, 48–52
 making fundamental changes 111–22
 and policies 117–18
 and politics 111–16
 and state origins 51
 and stewardship 118–22
 and urbicide 100–2
 as a worldview 95
 worldwide 60
 see also cities; economic development; supply issues

urban growth *see* economic development; urban demand
urban revolutions 80, *80*, 85
Urban Theory Lab-GSD 87
urban utopianism 88–9
Urbanocene 93, 95, 112
urbicide 102
Uruk, Mesopotamia 74, 81

V

van den Berg, M. 18
von Thünen, J.H. 54, 55, 87

W

wages 114
Wallerstein, I. 2, 40, 41, 45, 70, 92, 93, 100–1, 113, 114, 116
warlords 101, 102
water cycle 112
water supply 15
Weaver, W. 19
web of life 68–9
Weber, A. 60
Weber, M. 108, 109
Wenz, P.S. 69
Wheatley, P. 85
Williams, R. 60
work
 cheap work 114
 new work 23–5, *24*, 29, 31–3, 76, 103, 107–8, 110, 118, 119
 old work 23–5, *24*, 29, 31, 33, 36, 41, 76, 103, 119
World Bank 32
world-empires 82, 93, 101
world-systems perspective 40, 57, 58, 93, 100–1
 modern world-system 82–3, 92–3, 100–2, 106–7

Y

Yoffee, N. 50

Z

Zipp, S. 17, 18, 22, 31, 32, 33, 34, 36, 71